D0757855

Quantum Physics

Quantum Physics
A First Encounter

INTERFERENCE, ENTANGLEMENT, AND REALITY

Valerio Scarani
Department of Physics, University of Geneva, Switzerland

Translated by
Rachael Thew

OXFORD
UNIVERSITY PRESS

OXFORD
UNIVERSITY PRESS

Great Clarendon Street, Oxford OX2 6DP

Oxford University Press is a department of the University of Oxford.
It furthers the University's objective of excellence in research, scholarship,
and education by publishing worldwide in

Oxford New York

Auckland Cape Town Dar es Salaam Hong Kong Karachi
Kuala Lumpur Madrid Melbourne Mexico City Nairobi
New Delhi Shanghai Taipei Toronto

With offices in

Argentina Austria Brazil Chile Czech Republic France Greece
Guatemala Hungary Italy Japan Poland Portugal Singapore
South Korea Switzerland Thailand Turkey Ukraine Vietnam

Published in the United States
by Oxford University Press Inc., New York

Translation of Initiation à la Physique Quantique by Valerio Scarani
originally published in French by Vuibert, Paris 2003
Translated by Rachael Thew

First published in English 2006

British Library Cataloguing in Publication Data
Data available

Library of Congress Cataloging in Publication Data
Scarani, Valerio.
[Initiation à la physique quantique. English]
Quantum physics : a first encounter : interference, entanglement,
and reality / Valerio Scarani ; translated by Rachael Thew.
p. cm.
Includes bibliographical references and index.
ISBN-13: 978–0–19–857047–9 (acid-free paper)
ISBN-10: 0–19–857047–3 (acid-free paper)
1. Quantum theory. I. Title.
QC174.12.S3213 2006 530.12—dc22 2005020818

Typeset by Newgen Imaging Systems (P) Ltd., Chennai, India
Printed in Great Britain
on acid-free paper by
Clays Ltd., Bungay, Suffolk

ISBN 0–19–857047–3 (Hbk.) 978–0–19–857047–9

1

Contents

Part 2 Quantum Correlations

One Century of Quantum Revolutions

In 1900, Max Planck felt forced to accept the concept of the quantum in the exchange of energy between light and matter. Five years later, in 1905 – the miraculous year of which we celebrate the centenary – Albert Einstein enthusiastically decided to generalise the concept and to admit that light itself is composed of elementary quanta, later named photons, having both a specific energy and a specific momentum. He was of course aware of the difficulty of reconciling the image of a light beam being a flow of particles with the successful description developed during the nineteenth century of a wave, more precisely an electromagnetic wave. Only a wave can give rise to the phenomenon of interference and diffraction, as clearly shown by Thomas Young and Augustin Fresnel in the early 1800s. In 1909, Einstein delivered in Salzburg an amazing lecture[1] where he argued that we have to admit that light is both a wave and a particle, which may seem easy to say, but which is a challenge to our understanding. Since that time, quantum mechanics has not ceased to puzzle us. On the one hand, it is probably the most successful physical theory ever, allowing us to understand the microscopic world, notably the stability and properties of matter, and the manner in which light is emitted or absorbed by matter (for a long time light was the only way for us to get information about the intimate – microscopic, nature of matter). Quantum mechanics has also allowed physicists and engineers to invent and develop devices – such as the transistor and the laser – that have radically

[1] *The Collected Papers of Albert Einstein, vol.* 2, translated by A. Beck (Princeton University Press, Princeton NJ, 1989), pp. 379–398.

revolutionized our societies, by enabling the society of information and communication to emerge. However the basic concept on which this revolution is based, *wave-particle duality*, is still difficult to figure out using images developed from our macroscopic experience. In the first chapter of the volume on quantum mechanics of his famous 'Lectures on physics', written in the early 1960s, the great physicist Richard Feynman announces, about an experiment of interferences with electrons: '*In that chapter we shall tackle immediately the basic element of the mysterious behaviour in its most strange form. We choose to examine a phenomenon which is impossible,* absolutely *impossible, to explain in any classical way, and which has in it the heart of quantum mechanics. In reality, it contains the only mystery*'[2].

In fact, there is a another fundamental mystery in quantum mechanics. In 1935, Einstein and his colleagues Boris Podolsky and Nathan Rosen ('EPR') have discovered that the mathematical formalism of quantum mechanics allows for a remarkable quantum state of two particles, which Schrödinger also considered at about the same time, and named 'an entangled state'. The correlations between the two entangled particles are predicted to be so strong, that Einstein and his collaborators found themselves entitled to challenge the mainstream interpretation of quantum mechanics developed by Niels Bohr and his collaborators (the so-called Copenhagen interpretation), where the predictions of quantum mechanics are probabilistic in their very essence. When you consider these correlations between separated particles, EPR argued, it is hard to escape the conclusion that quantum mechanics has to be completed, that there is an underlying level of description where these particles have supplementary properties, not appearing in standard quantum mechanics[3]. Niels Bohr immediately opposed

[2] R.P. Feynman, R.B. Leighton, and M Sands, *The Feynman Lectures on Physics: Quantum Mechanics,* section 1.1 (Addison-Wesley, Redwood City, 1989).

[3] Their reasoning was exactly the same as biologists observing strong correlations among physical or medical features of twin brothers (or sisters) and concluding, before the observation of chromosomes with an electron microscope, that twin brothers must have identical features that one can call genes.

that conclusion. He apparently felt that the formalism of quantum mechanics could not be completed without becoming inconsistent and collapsing. The debate on this question between the two giants of physics, lasted until their deaths, twenty years later, but it is fair to say that it attracted little attention from most other physicists. On the one hand, mainstream physicists were successfully using quantum mechanics to understand more and more subtle phenomenon, and to invent new devices. On the other hand, there was no disagreement between Einstein and Bohr on the results of the quantum mechanical calculations, only on the interpretation of these results. The theory was thus believed to be safely utilizable for all practical purposes.

Thirty years later, John Bell showed that it was not the case. In 1964, in a now famous short paper[4], (published in an obscure journal that disappeared after four issues), Bell showed that if you take seriously the point of view of Einstein and complete quantum mechanics accordingly, then you face a quantitative contradiction with some predictions of quantum mechanics. It is no longer a matter of interpretation, but now an open question of whether nature behaves according to quantum mechanics or to Einstein's world view. Considering the overwhelming success of quantum mechanics, it may seem that the answer was obvious, but it was not so. As a matter of fact, situations in which the conflict arises are very rare, actually they are only EPR type situations, with entangled particles, and even in such situations one has to make very specific measurements which had never been done. But at the time of its publication, Bell's work met little interest, not to say open hostility, and it is only after a small group of people led by John Clauser and Abner Shimony proposed a practical experimental scheme, and when clear cut results became available, that the importance of Bell's work was recognized. Progressively, it became clear that *the strangeness of entanglement was something different from wave-particle*

[4] J.S. Bell, *On the Einstein Podolsky Rosen paradox* (1964), reproduced in J.S. Bell, *Speakable and unspeakable in quantum mechanics*, 2nd edition (Cambridge University Press, 2004).

duality. To illustrate the evolution of 'mainstream physicists', it is illuminating to quote again Richard Feynman. In his 'lectures on physics', in the early 1960s, this is what he had written about the EPR argument, reflecting (and influencing) the opinion of the immense majority of physicists: '*This point was never accepted by Einstein... it became known as the "Einstein-Podolsky-Rosen paradox". But when the situation is described as we have done it here, there doesn't seem to be any paradox at all.*'[5] Twenty years later, however, Feynman seemed to have changed his mind, since he wrote, about the conflict between quantum mechanics and Bell's type inequalities: '*We always have had a great deal of difficulty in understanding the world view that quantum mechanics represents... I have entertained myself always by squeezing the difficulty of quantum mechanics into a smaller and smaller place, so as to get more and more worried about that particular item. It seems to be almost ridiculous that you can squeeze it to a numerical question that one thing is bigger than another. But there you are — it is bigger.*'[6] The weirdness of entanglement could not be reduced to the quantum mystery of wave-particle duality.

To emphasize the radical difference between the conceptual problems raised by wave-particle duality, on the one hand, and entanglement on the other hand, I have proposed[7] to name '*the second quantum revolution*'[8] what has been happening starting in the 1960s, events that have been without any doubt provoked by Bell's reanalysis of the EPR argument. Actually, the second quantum revolution is based not only on entanglement, but also on the development by physicists, also beginning in the 1960s, of methods for controlling, trapping, observing single microscopic objects such as electrons, atoms, ions, photons, and molecules, or even

---

[5] R.P. Feynman, section 18.3 of ref. 1.

[6] R.P. Feynman, *International Journal of Theoretical Physics*, **21**, 467 (1982).

[7] A. Aspect, introduction to *Speakable and Unspeakable in Quantum Mechanics*, see ref. 3.

[8] Daniel Kleppner, a famous professor of Physics at the Massachusetts Institute of Technology, suggested recently the alternative name 'a new quantum age' rather than of 'a second quantum revolution', which I like equally well, perhaps more.

devices as Josephson junctions. The application of the methods of quantum mechanics, a priori well suited to the description of large ensembles, to describe the behaviour of single quantum objects, gave rise to an interesting clarification. Here again, as in the case of entanglement, all the theoretical tools were already available in the general formalism of quantum mechanics, but it took some effort for physicists to find clear ways of using them in these new situations. My claim is that these two conceptual and experimental advances of the 1960s – realising how far-reaching entanglement is, and mastering experimentally and conceptually single microscopic objects – are at the root of the new quantum revolution, of which we do not know the outcome. Will that new conceptual revolution trigger in turn a new technological revolution, that might change our society, just as the first quantum revolution did it? It is too early to say, but we have already seen some advances in that direction. Since the 1980s, it has been suggested that entanglement, being an entirely new physical concept, might enable the development of entirely new ways of processing information: this is the emerging field of *quantum information*, in which computer scientists, mathematicians, theoretical physicists, and experimentalists work together. The first branch of quantum information, *quantum cryptography*, has already been developed to a point where small 'start up' companies offer systems of communication where the privacy is guaranteed by the very laws of quantum physics. The second branch of quantum information, *quantum computing*, is certainly less advanced, but a whole community of experimentalists works at achieving 'entanglement on demand' of more and more quantum bits (named 'qubits'), while theorists propose new algorithms for using a would-be quantum computer. In June 2005, the state of the art was seven entangled qubits. There is a long way to go, fighting the problem of decoherence, before building a quantum computer that would be able to achieve tasks inaccessible to our biggest computers, such as factoring large numbers. Nobody can tell if useful quantum computers will exist one day, but this field of research has attracted outstanding teams all over the world.

The book of Valerio Scarani presents on an equal footing the first and the second quantum revolution. He does not hide that the concepts at play are difficult to understand: in fact, even professional physicists admit that although they know how to use quantum mechanics, they have difficulty integrating it into their image of the world. This difficulty should not be hidden, it is part of human culture, which grows on the accumulation of achievements in all domains of human thinking, including science. It would be a tragedy were the general public to renounce awareness of science advancement, and the book of Valerio Scarani makes every effort to present the conceptual revolutions of quantum mechanics to the general public. It seems to me that he reaches his target. The reader who accompanies him (along with the students of the philosophy class at Saint Michel college of Fribourg) to this destination will get a much better idea of the strange world of quantum physics, which is already familiar to physicists, and also becoming more familiar to engineers working on nanotechnologies. Last, but not least, the book is written in a light and pleasant style, and shows that discovering the mysteries of quantum mechanics does not have to be boring, but can be exciting.

Orsay, July 2005
Alain Aspect

Prologue

California Institute of Technology (Caltech), 1984. Alain Aspect, a young French physicist, is giving a seminar about his recent experiments. Richard Feynman, Nobel-Prize winner in 1965, sits in the audience. Nothing but a commonplace event in the academic life of one of the most renowned research institutes in the world, a weekly seminar, similar to the one heard seven days before and to the one scheduled for seven days later. Still, commonplace events may acquire a very different status with the passage of time. Feynman is one of the great figures of quantum physics, and also one of its best teachers. In his lectures for undergraduate students[1], published the same year as his Nobel Prize is awarded, he deals with 'everything' that you can expect to find in an introductory course – everything about physics. He does not indulge in the mathematics he masters so well, rather, he stresses phenomena. In the same year as Aspect's seminar, Feynman is going to deliver his series of three lectures for a wider audience, a masterpiece in the communication of science[2]. The speaker to whom he is listening, Aspect, is not such a giant. He has just received his Ph.D., but the experiments he has performed are rapidly being recognized as a cornerstone of physics.

Aspect has finished his talk; the time has come for questions from the audience. Feynman raises his hand and asks to see one of the transparencies again[3]. Aspect puts it on the overhead projector. 'Having these elements at your disposal, couldn't you perform an experiment I have wanted to see for so many years?' Aspect has presented one of the first pieces of evidence of the 'second quantum revolution'[4]; Feynman's interest goes back to the first quantum revolution. The American physicist has realized that an experiment could be performed that would show immediately, strikingly, the difference between classical and quantum physics, without having to rely on indirect evidence, huge experimental apparatuses and some mathematical background. 'Yes, it can be done' is Aspect's answer . . . he had actually already had the idea: at that very moment, while he is lecturing in California, a student of his in Paris is assembling everything necessary to perform the experiment that Feynman is waiting for[5]. A few more questions on the foundations of quantum mechanics and on technical details, and the audience disperses. Nobody imagines that the event they have just witnessed signifies the twilight of the traditional approach to quantum physics and the dawn of a new vision[6].

The experiment suggested by Feynman and the experiments described by Aspect in his talk are the main themes of this book (Part One and Part Two, respectively). These phenomena strongly challenge the vision of nature based on our daily experience. Confronted by them, we are astonished, surprised, fascinated and simultaneously disturbed – a rather suitable position for starting a philosophical reflection[7]. Indeed, it is well known that quantum physics gives rise to important implications for our worldview. There is no agreement among physicists as to whether one can speak of quantum physics without adhering to an interpretation. Throughout most of this book, following the spirit of Feynman, the accent is on description of phenomena, whence a single interpretational principle ('indistinguishability principle') is stated – this can be seen as the most minimalist interpretation, the principle being intended as a guide to classification rather than as a deep metaphysical statement. A review of more involved (thus, in some sense, more interesting) interpretations is given at the end of the book, once the phenomena have been understood.

This text is based on a series of lectures given to various audiences. Once this pedagogical approach is set in black and white, it becomes impossible to shorten what sounds boring to the audience or linger over that which fascinates them. It is all the more necessary to warn the reader that this text does not go from the simpler to the more complicated. My lectures dealt principally with what is found in Chapters 1, 6 and 9, which therefore form the skeleton of the whole book. The discussion of the Bell theorem (first section Chapter 7) is certainly the most arduous section.

This initiation into quantum physics has benefited from the clear vision of quantum physics first conveyed to me by François Reuse and Antoine Suarez in my university years, then confirmed and deepened through close contact with Nicolas Gisin and many other physicists of quantum information – among whom I would like to single out Sandu Popescu. The original French edition has been possible thanks to the encouragement of Jean-Marc Lévy-Leblond; it has benefited from feedback from so many sources that I can't possibly mention them all here. The present English edition owes a lot to Alain Aspect, Mark Fox and Sonke Adlung for the editorial process, to Rachael Thew who made the translation, as well as to the criticism of an anonymous reviewer.

Part 1
Quantum Interference

·1·
At the heart of the problem

1.1 Fribourg, May 1997

'When all is said and done, physics is only a description, not an explanation'. I have uttered similar statements many times myself in informal discussions, when my interlocutor happened to slide into an excessive praise of experimental science and an excessive contempt of other forms of human knowledge. In the mouth of Jean-Paul Fragnière, a friend and philosophy teacher who lives in the small Swiss town of Fribourg, this sentence sounds, however, the expression of a different and distant line of thought... of a rivalry between human and natural sciences that is still latent in the French-speaking world, a relic of the arrogance of Positivism.

We are in the massive building of Saint-Michel College – a sixteenth century Jesuit establishment, a public school for several decades; and though Jean-Paul's classroom is a modernization of what was the granary, history weighs heavily here. For the first generations of young people educated in this school, heliocentrism was one of the hot topics; now Jean-Paul has invited me to speak to his students about quantum physics. He is concerned by the theme, and what philosophy teacher wouldn't be when their students come to consult them after having read or heard that quantum physics has abolished the notion of causality? While the students arrive one after another and greet the two of us timidly, he explains that the question of the interpretation of modern science interests him very much, although without troubling him particularly because 'physics is only a description, not an explanation.' At the time, I didn't react – fortunately, I was not required to.

When the audience is assembled, Jean-Paul introduces me with the hesitation of a friend who discovers that he does not know my academic qualifications. We go straight to the heart of quantum physics – beginning with the experiment Feynman asked Aspect to perform[8].

1.2 First observations

1.2.1 Semi-transparent mirrors

If someone had entered the classroom a few minutes after the start of the lecture, they would have seen that the gaze of the whole audience was directed towards the window. An understandable boredom? In this case, quite the contrary – the sign of an active interest. I had mentioned the windowpane as a most ordinary example of a *semi-transparent mirror*. Those looking in from the outside could see us behind the glass, but we can also see our own image reflected by the pane. That means that some of the light that we emit inside the room is *transmitted* through the glass, and some is *reflected* by it.

Generally speaking, a semi-transparent mirror is an object that splits a beam into two parts, which is why we also use the name *beam splitter*. The beam can be a ray of light, as in the example of the windowpane, but at least in principle, we can construct splitters for beams of particles[9]: neutrons, atoms, electrons . . . In Chapter 3, we describe the neutron beam splitter. For the moment, we are going to remain at a very general level, since the phenomena that we are going to describe are produced by all particles. Let's assume then that we have at our disposal: (1) a *source* that emits a particle beam; (2) some *beam splitters*; and (3) some *detectors* for those same particles. A detector here is simply a measuring device that allows us to count the particles.

1.2.2 First experiment

We fire the particles, one by one, at a semi-transparent mirror, and we count how many of them are transmitted (T) and how many are

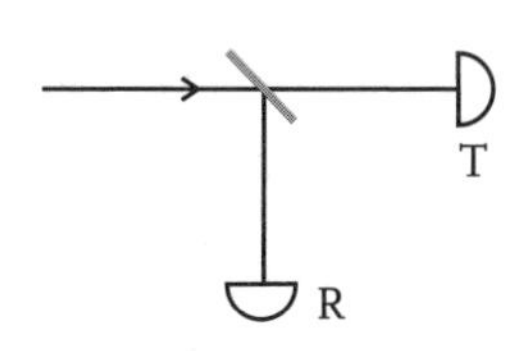

Fig. 1.1 Apparatus illustrating the function of a semi-transparent mirror (beam splitter).

reflected (R). This experiment is illustrated in **Fig. 1.1**. After having fired a large number of particles, we can make two observations:

First, the two detectors are never activated at the same time, which indicates that a particle arriving at a beam splitter is not divided, *it is either transmitted or reflected* – incidentally, this is the feature that enthralled Feynman: performed with a light beam, this experiment clearly demonstrates the particle nature of light, superseding the ingenious but indirect path that Einstein had to take in 1905.

Second, if we count how many particles have taken each route, we notice that half of them are transmitted and the other half are reflected. This claim must be understood just as it would be as regards a game of heads or tails – if we toss a coin a large number of times, it is impossible to predict each result, but we know that in the end the number of 'heads' and the number of 'tails' will be just about the same. We refer to this situation by saying that *the probability of an event is* 50% or 1/2. Therefore, expressed in precise terms, the result of this first experiment is: the probability that a particle is transmitted through the beam splitter is equal to the probability that the particle is reflected. This amounts to saying that the two probabilities are 50%, as the total must be 100%.

Some of the students exchange glances, then interrupt me. They declare themselves to be rather surprised at the comparison between particles and the tossing of a coin – are particles, physical objects, able to behave *at random*? It is delightful to hear this pertinent question, which has caused so much ink to flow during the past century. Jean-Paul delights in it too, and I see that he is preparing to ask his students what they understand by the word *random*, one of those words used too often lightly, which a professor of philosophy cannot let pass. Nevertheless, I must play for time and avoid getting trapped in a discussion on randomness and determinism at this early stage: randomness is a key ingredient of quantum physics, but quantum physics is more than just a question of randomness. I ask then that for the moment they accept that we use a language of probability to describe the behaviour of a particle that encounters a beam splitter. The topic of (in)determinism will be tackled among the interpretational issues, but we must first become acquainted with the phenomena.

1.2.3 Second experiment

In order to familiarize ourselves with beam splitters, it is helpful to complicate the experiment a little, as in **Fig. 1.2**. This time, after each output from the first beam splitter, we put another beam splitter. Thus, the apparatus has four outputs – the particle may be transmitted twice (TT), transmitted at the first beam splitter and reflected at the second (TR), reflected at the first and transmitted at the second (RT) or reflected both times (RR). What probability do we expect for each of the outputs?

We cannot respond in advance to this question. Let's suppose that each particle carries an 'instruction', so that each time it encounters a beam splitter it is either definitely transmitted or definitely reflected. In this case, we will find half the particles at TT and the other half at RR, with no particles at TR or RT.

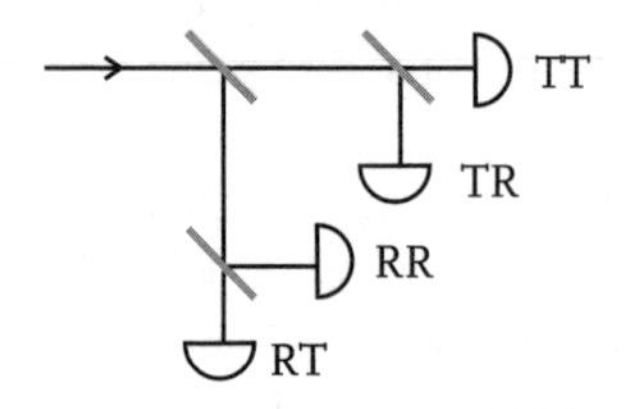

Fig. 1.2 Apparatus with three beam splitters, defining four paths.

But this is not what we observe on firing a large number of particles into the apparatus. In fact, as always as regards probabilities, we find 25% of the particles at each output. This again resembles the game of heads or tails – in throwing the same coin twice in succession, we expect to find the four cases heads-heads, heads-tails, tails-heads, tails-tails with the same probability. Apart from the recurrent issue of randomness, which we agreed to leave aside for a while, this observation is certainly not surprising.

1.3 Interferometry

1.3.1 The initial observation

The next apparatus through which we are going to pass the particle is illustrated in **Fig. 1.3**. In this apparatus there are two perfect mirrors that reflect all particles, thanks to which we can direct the two output paths of the first beam splitter to the same second beam splitter. In this way, one of the outputs of the second beam splitter corresponds to the paths RT or TR, and the other to the paths TT or RR.

Understanding experiment 2, this new setup does not seem to pose any problems – since on the RT path we have found 25% of the particles, and the same percentage on the TR path, then at the output at *RT or TR* of our current apparatus we expect to find 25% + 25% = 50% of the particles. Naturally, the other 50%

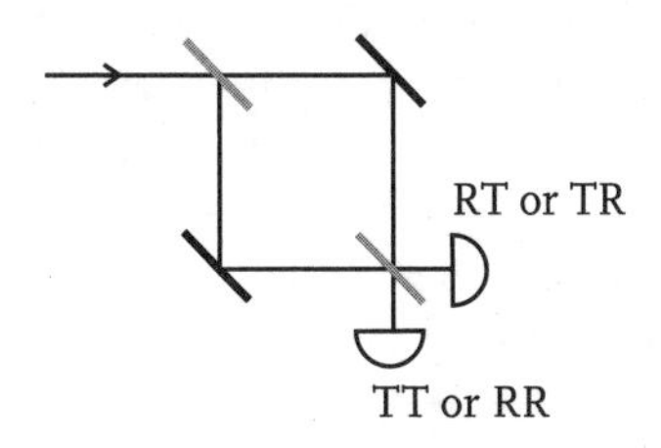

Fig. 1.3 Balanced Mach–Zehnder interferometer.

are to be found at the output *TT or RR*. Let's summarize this conclusion:

PREDICTION: 50% of the particles arrive at the output *RT or TR*, the other 50% arrive at the output *TT or RR*.

And yet that is not what is observed! In fact, the observation is dramatic:

OBSERVATION: all of the particles are found at the output *RT or TR*.

We have doubtless missed something, but what? Let's calmly review the evidence we had collected. Experiment 2 is irrefutable: a quarter of the particles were definitely at the TT output, and another quarter at RR. Our beam splitters therefore function correctly. Moreover, we definitely sent one particle after the other so it is impossible that, through an unfortunate coincidence, two or more particles encountered each other at the second beam splitter and that the output was dictated by an unwanted collision. Finally, experiment 1 has shown us that the particles were indivisible – we have never detected a half-particle at any detector, each time it was one and only one detector that was activated. Everything seems in order, and yet nature behaves in an unexpected manner in experiment 3. We are touching on the heart of quantum physics, and the surprises are not over yet.

1.3.2 More and more surprising

In the apparatus in **Fig. 1.3**, the paths that the particle can take are the same length. Let's change the length of one of the two paths (**Fig. 1.4**). As soon as the lengths of the paths are different, some particles (a small number, if the difference in length is small) arrive at the *TT or RR* output. The greater the difference in the length of the paths, the more particles are found at the *TT or RR* output. When the two paths differ by a certain length *L*, *all* of the particles are found at the *TT or RR* output, and no particles are found at *RT or TR*. If we were to continue to increase the length, the inverse effect would be produced and for a difference of 2*L* all of the particles

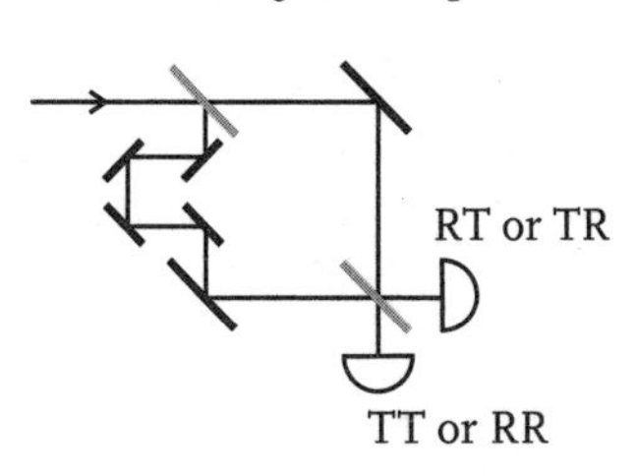

Fig. 1.4 Slightly unbalanced Mach–Zehnder interferometer.

would arrive at the *RT or TR* output, exactly as when the two paths are the same. We could recommence the cycle, but let's leave it there[10].

The importance of this new piece of information can be clarified by formulating a question: how does it work, that in changing *only one* of the two paths, we manage to change the behaviour of *all* of the particles? How is it that the particles that travel by the path that we have not modified can know about the modification? Yet this is exactly what we observe, this is how nature behaves. We must conclude that each particle is 'informed' about all of the paths that it could take, without being actually 'divided' into two parts: Experiment 1 tells us that if we look at which path the particle takes, we would find it on either one or the other, never on both.

We can investigate this point further: suppose that indeed we find a way of measuring whether the particle has been transmitted or reflected at the first beam splitter – that is, of measuring which path the particle has taken. The results of experiment 1 are certainly confirmed – we find the particle sometimes on one path, sometimes on the other, with equal probability. On the other hand, it is the results of experiment 3 that change completely – if we know the path followed by each particle, then at detection half the particles are found at *TT or RR*, the other half at *RT or TR*, whatever the difference in length of the two paths. To put it plainly, if we try to know by which path the particle is travelling, we completely lose the surprising effects of apparatus 3 – the particles behave according to our intuitive prediction.

This 'bizarre' behaviour of the particles has a name. We refer to *single-particle interference*, or the fact that the particle *interferes with itself.* The reason for this name will be clarified in the next chapter. Apparatus 3 is called the *Mach–Zehnder interferometer.*

1.3.3 Indistinguishability principle

I have introduced several quantum phenomena in this first lecture; more (and weirder) are to come in the second lecture, scheduled in the same classroom for the following week. Only at the end of the second lecture will the students and the reader have a solid enough foundation from which to embark upon a serious tour of the interpretations of quantum physics. Remember that phenomena are indisputable, while syntheses, descriptions and explanations are not: apples fall, whether because their natural state is 'down', or because the earth exerts a force at a distance called gravitational attraction, or because of a deformation of space-time. But from the apple we also learn, first, that not all of the descriptions are equivalent, because some of them grasp more of reality than others; secondly, that science has made real breakthroughs when it has moved away from the redaction of a catalogue of phenomena, towards a synthesis[11]. The students themselves would appreciate leaving for home with some 'explanation', however provisional.

Actually, one can state a principle that *describes* the conditions under which a particle interferes with itself. Admittedly, to state a principle is not an entirely satisfactory solution, and definitely not a valid explanation, but at least it allows a synthetic presentation of the experiments, and thus constitutes the least 'committed' interpretation, the safest step. This principle, called *the indistinguishability principle*[12], can be expressed like this:

Interference appears when a particle can take several paths in order to arrive at the same detector, and the paths are indistinguishable after detection.

Let's put the principle to work on the phenomena we have already described. In apparatus 1 and 2, there is only one path leading to each detector; consequently, when a particle is detected we know exactly which path it must have taken. It is a situation of distinguishability and no interference effect is apparent. In apparatus 3 and 4, on the other hand, when a particle is detected after the second beam splitter, we have no way of knowing by which path it arrived, since two paths are possible. These two paths are therefore indistinguishable, and the effects of interference are present. Finally, we have seen that the interference disappears again if we detect the presence of the particle on one of the two paths. More generally, it disappears if the particle leaves an imprint of its passage on the path, because in this case the imprint on the path destroys the indistinguishability of the two paths.

1.4 End of the first class

It is enough for the first lecture: 'human kind cannot bear very much reality'[13], all the more after a full school day[14]. Time to arrange next week's rendezvous with them, and the students are leaving scarcely saying their goodbyes, with the apparent ingratitude of their age. Jean-Paul and I close the classroom door, descend the narrow stairwell, walk down the imposing corridor between portraits of former rectors, and emerge in the main courtyard of the college: back into real life – or rather, the reality of the particles' behaviour forcing some modesty on us: back into *ordinary* life.

·2·
Let's step back

2.1 Questions and properties

2.1.1 Indistinguishable cars

Leaving Saint-Michel, I take my leave of Jean-Paul who is going home, and I turn left to walk down the College Steps, a covered lane that descends into the heart of Fribourg. There, in front of the cathedral, three busy roads meet to form a roundabout. I continue my walk along the south side of the cathedral. Cars continue to stream down the road. Now that the roundabout is no longer in view, I cannot know by which route each car has entered it. In order to arrive alongside me, as it does now, this car could have taken three possible paths. I have put myself, in relation to the cars, in a situation of indistinguishability. Well then, why is the road network not a quantum interferometer? The answer is a matter of common sense – the indistinguishability of cars is only apparent, it is due to an ignorance that can easily be overcome. In order to obtain the information about the path by which each car entered the roundabout, it would suffice for me to go back a few paces or to ask the driver – neither of these steps would modify what the destination of the car will be. In quantum physics, however, the knowledge of the path taken by a particle may change its fate dramatically . . .

2.1.2 Everyday questions

I now propose to the reader a small exercise. Let's consider all of the cars that are circulating on the roads of the municipality of Fribourg at this moment. We have a *set of cars*[15]. In posing a question to which

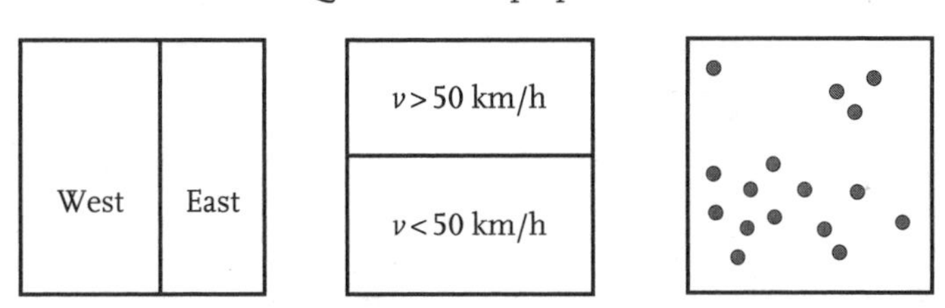

Fig. 2.1 Definition of classical binary properties on a set of cars. The grey dots in the third square represent the elements of the set for which the response to question 3 is *yes*.

the response can be *yes* or *no*, we are going to divide the set into *two subsets*. For example, I pose question 1: is the car located to the east of the river Sarine that bisects the town? This question allows the separation of the set of cars into two subsets – the cars for which the response to the question is *yes* (therefore those that are located to the east of the Sarine), and those for which the response is *no* (and that are therefore located to the west, or on a bridge). Likewise, I can pose question 2: is the car obeying the 50 km/h speed limit? This question also introduces a separation into two subsets, a separation that is, in principle, different from the preceding one. Naturally, I can pose more elaborate questions, as in question 3: has the car been in the municipality of Fribourg for more than a quarter of an hour?

We can schematically represent the subsets associated with these questions as we have done in **Fig. 2.1**. Once drawn like this, one sees quite immediately that the notion of a set implies a richer structure: one can combine the questions, that is, one can ask if the car is located to the east of the Sarine *and* if it is obeying the speed limit. The set that describes the *no* response to this question is the intersection of the two sets corresponding to the *no* response at questions 1 and 2 (**Fig. 2.2a**). With a little practice, the reader can manage to prove to themselves that the *and* of the logic corresponds to an intersection of sets, the *or* to a union of sets, the *not* associates a set with its complement, the *implication* (if A then B) is expressed in the inclusion of one set within another, and so on.

(a)

West & $v>50$	East & $v>50$
West & $v<50$	East & $v<50$

(b)

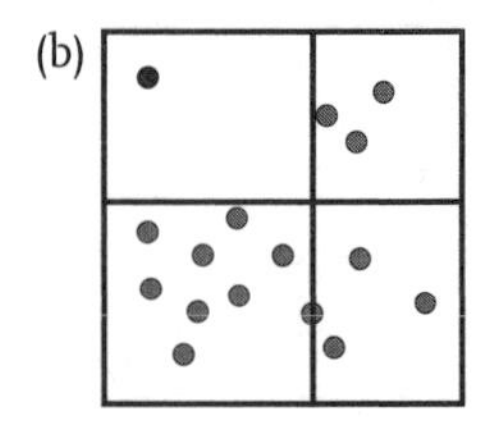

Fig. 2.2 Definition of combined classical properties (AND) by the intersection of the sets.

Ordinary logic can be formalized with the tools of set theory. But why? This correspondence between our manner of reasoning and a mathematical structure must have a reason.

Indeed it does have one – we suppose that we can ask all of the questions that we want to ask, one after another, about cars, and that all of the answers will be meaningful. Each question is going to divide the set of cars into two parts, if we ask a sufficient number of questions we will have finer and finer divisions, and in the end each car will be characterized by a list of *yes* and *no*. In our diagrams, for example, there is only one car located to the west of the Sarine that is not obeying the speed limit and that has been within the municipality for more than a quarter of an hour (the black spot in **Fig. 2.2b**). This car is characterized by the following data: question 1: no; question 2: no; question 3: yes.

2.1.3 Quantum questions

Let's go back now to our quantum interferometry from **Fig. 1.3**. For simplicity, we will refer to the detectors that count the particles at the *RT or TR* output and the *TT or RR* output as D_1 and D_2, respectively. Instead of cars, we have particles. Let's now consider the two following questions:

1. Has the particle taken path T to the first beam splitter?
2. Has the particle been detected at D_1?

These are two questions to which we can answer *yes* or *no*, therefore we should be able to repeat the same exercise as we have just done with the sets of cars. For example, we should be able to define the set of all of the particles that take path T (which I shall call S_1) and the set of all of the particles detected at D_1 (which I will call S_2), and we should be able to identify the particles that have travelled by path T *and* have been detected at D_1, by taking the intersection I at S_1 and S_2. This should be correct...

However, if we refer to the experiments presented in the preceding chapter, the analysis of questions 1 and 2 becomes problematic. In effect, we can consider two cases:

Case I: if we do not modify the experiment, we know that the response to question 2 is always *yes*, because all of the particles arrive at the D_1 output. But in this experimental configuration, we *cannot respond* to question 1.

Case II: if we modify the experiment by inserting detectors on the paths R and T so as to be able to respond to question 1, we change the response to question 2 – this time, only half the particles would give a *yes* response. The response to one question consequently depends on the other questions that we pose!

It is obvious that case I and case II are incompatible. We have already noted this in the preceding chapter, when we were forced to admit that each particle explores both paths R and T. Let's see how the problem arises here.

We have said that in case I we cannot give an experimental response to question 1. This means that it is impossible for us to know by which path the particle has travelled, without modifying the result. But can we at least try to *guess* the response to question 1? In other words, does question 1 have a 'hidden response'? Not even: let's prove this impossibility by *reductio ad absurdum*. Let's then assume that this hidden response exists, that is to say that a certain percentage *p* of particles actually take path R. Let's consider then the particles that have taken path T – we do not know *which* particles have taken this path, but we are assuming that this set *exists*. How do the particles of this set behave on arriving at the second beam splitter? In order to safeguard the coherence of the observations, *they must behave according to experiment* 1 of the preceding chapter. In effect, we are assuming that these particles arrive at the beam splitter by a clearly defined path. Consequently, we predict that half the particles of this set will arrive at D_1, the other half at D_2. We can apply the same reasoning to the particles that we assume have taken path R – among these also, half will arrive at D_1, the other half at D_2. Therefore, *whatever the percentage p that we assume,* the result at detection is: half of the particles at D_1, the other half at D_2. No other result is possible for case I under the hypothesis that each particle has taken either path T or path R. But, as we have seen, the experiment gives a different result. We are forced to conclude that question 1 *does not really have a response* in case I. We can devise and carry out an experiment for which question 1 gives a response (that is case II), but in doing so we modify the results of detection.

In summary: in Chapter 1, we arrived at a surprising phenomenological conclusion, ascertaining the fact that the particles explore all of the possible paths while manifesting themselves on only one path when their position is measured. We have just derived here one of the main reasons for our surprise – the physical properties of quantum systems, contrary to the physical properties of cars or other everyday objects, are not bound to each other according to the rules of set theory. In other words, we cannot combine as we please the questions on the physical properties actually possessed by a quantum system[16].

The epistemological analysis that we have made clearly opens important interpretational issues, on which Chapter 9 will hopefully cast more light. Now, after having stepped back conceptually from the phenomena of Chapter 1, we must step back factually, into history.

2.2 Waves and corpuscles

Physicists did not create quantum physics from start to finish. In a way, it was given to them, and in any case it was unexpected. I shall leave it to others to take care of deciding whether it was a matter of a deserved reward for services given to knowledge, a maternal nudge in the right direction from nature in order to prevent us from straying too far from the reality of things, or a completely fortuitous combination of circumstances. The brief look at history that follows is extremely schematic, in the literal sense of the word. I am going to select data from history that is most suitable for the content of this book[17]. We are going therefore to cover the genesis of the notion of the particle and that of interferometry experiments.

2.2.1 Brief history of particles

These days, every good school child recites that 'matter is composed of atoms', the word 'molecule' appears in chemistry, biology and physics courses, and physicists, who in the rings of huge accelerators have already shown a menagerie of particles, strive to track particles arriving from space. The existence of 'particles' has entered into the collective consciousness.

Yet, scarcely a hundred years ago, on the subject of 'particles', one of the most tragic pages in the history of science was written. The Austrian physicist Ludwig Boltzmann, who had devoted years of effort to gathering arguments supporting atomism, was driven to depression and eventually suicide by the attitude of the scientific community, deaf to his ideas. Ironically, it would have sufficed for Boltzmann to survive another three short years in order to see official science doing a complete about-face, accepting atomism and so fulfilling one of his goals.

The idea that material reality is not 'continuous' (a kind of jelly, on all scales) but formed from 'elementary building blocks' (the 'atoms') had already been proposed by some ancient Greek thinkers. In order to better grasp the surprising side of quantum physics (which is the aim of this text), it is worth the effort of devoting a few paragraphs to the motivation that revived the trend of atomism in the nineteenth century[18].

This nineteenth century was a century of positivist optimism – science, and physics in particular, achieved a growing number of successes. The declarations of faith in the omnipotence of science, which one day would be able to answer all of our questions, are multiplying, some people unhesitatingly claim that the solution to all of the problems of knowledge is close at hand. It would, however, be a mistake to believe that the physics of the nineteenth century, which aroused such enthusiasm, was a monolithic knowledge, compact and neatly packaged. On the contrary, physics is presented as being divided into disciplines, each with its own history, each having only the weakest of links to the other branches.

The king of all disciplines is without doubt *mechanics*, the study of the motion of bodies. The accurate predictions of mechanics are innumerable, its mathematical tools are at once powerful and elegant, and it merits the apposition of the most laudatory adjective there is in the spirit of that age: we speak of *rational* mechanics. *Thermodynamics* is relatively recent in the nineteenth century: the study of exchanges of energy, of heat, of temperature, has had considerable growth since the invention of the steam engine. *Fluid mechanics* (flow of water, of a gas...) was established on a less solid basis than the mechanics of solid bodies: the turbulence of a watercourse is far more complex to describe than the motion of the Earth around the Sun. That said, and despite the sometimes brutal approximations that one makes in order to solve its equations, the predictions of this branch are good too. Finally, the great novelty of the nineteenth century comes from the detailed description of certain phenomena with which we have been partially acquainted since antiquity, *electrical* and *magnetic* phenomena. This part of physics will only be completed at the beginning of the twentieth century and will bear the name *electromagnetism*. The attentive reader has certainly noticed the absence from this list of a discipline as ancient as mechanics, *optics*, the study of light. Light has a special status in the physics of the nineteenth century and we will devote the next section to it.

Let's summarise: mechanics, thermodynamics, fluid mechanics, electricity and magnetism. This list corresponds to the physics programme in high school and the first years of university – the reader probably knows from experience how these disciplines can appear without any link between one and the others. In this context, the hypothesis of atomism is presented as a *unifying vision*. Atomism proposes to reduce all physical phenomena of matter (thermal, electrical, magnetic, turbulence phenomena) to mechanical phenomena – the manner of the motion of atoms. For example, an electrical current will be described as the displacement of particles carrying an 'electrical charge', the temperature of a gas will be associated with the average speed of the particles that form the gas. On the scale of particles, we would like there to be only intrinsic properties (such as mass and electrical charge) and motion.

The atomic hypothesis proves to be fruitful and, as we have mentioned above, eventually imposes itself at the beginning of the twentieth century, not without polemics, difficulties or drama. And there lies the big surprise: the mechanics of these 'atoms', of these 'particles', this mechanics that is supposed to unify all phenomena, *is not the usual mechanics* mastered two centuries before! In their motion, in their energetic properties, the particles behave according to new laws. Of these unexpected laws, the reader has already had an insight in the preceding chapter, but it is necessary to underline that the clarifying notion of indistinguishability, on which we have based our description, did not leap immediately to the attention of physicists. This is why the physicists at the beginning of the twentieth century did not forge the name *indistinguishability mechanics*, but began to speak about *wave mechanics* or *quantum mechanics*. It is to these two adjectives that the following two sections are devoted.

2.2.2 Brief history of interference

We want to look into the first of the two historical names for the mechanics of atoms – *wave mechanics*. A short introduction to the notion of the wave is certainly not unnecessary at this stage. Contrary to what we might think, the word *wave* in physics refers to a rather *abstract* idea. The reader should make a list of the different types of waves of which they have heard: waves of water, sound waves, radio waves, microwaves... Sound waves and radio waves are very different from each other, that is well known, and moreover, if radio waves could be heard by the human ear, our daily lives would take place amid an unbearable din. In physics courses, we give the name 'wave' as much to the vibration of the string of a violin as to the sound that this vibration produces. In summary, the concept of the 'wave' is a concept that *allows the description of a class of phenomena*. It is therefore even more abstract than that of the 'particle', which initially evokes an object.

The history of physics has been marked during several centuries by the question of knowing whether *light* was a beam of corpuscles or, actually, a wave phenomenon. We are touching on the domain of classical physics that we had left to one side in the preceding section, namely optics. Let's put ourselves in the seventeenth and eighteenth centuries, an epoch of great debates between adherents of the corpuscular description (Newton, followed by the English school) and those of the wave description (some continental scholars, following the Dutchman Huygens). The experiment allowing the abrupt resolution between these two descriptions was proposed and carried out in the first years of the nineteenth century by the English scholar Thomas Young (the same man who had reported some partial successes in the deciphering of Egyptian hieroglyphs). The wave description of light gained the upper hand. What was this experiment that was so clear? What principle of physics came into play that distinctly separates the wave and corpuscle phenomena? The reader has already encountered the key word several times in this book – Young's experiment was an *interferometry* experiment!

In *Young's interferometer*, the light encounters an opaque barrier in which two slits have been pierced (**Fig. 2.3**). On the other side of the barrier, the intensity of the light is visualized on a screen – the modern reader might think of a photographic plate. The principle is the same as that for the Mach–Zehnder interferometer that we described in Chapter 1 – each 'point' of the screen plays the role of detector, every point on the screen may be reached by *two indistinguishable paths*, and the *difference in length* of these paths determines the intensity of the light on that point[19]. The correspondence in the behaviours of the two apparatus according to the difference in the paths is given by the following rule: (i) the differences in length over which all of the light is detected at *RT or TR* in

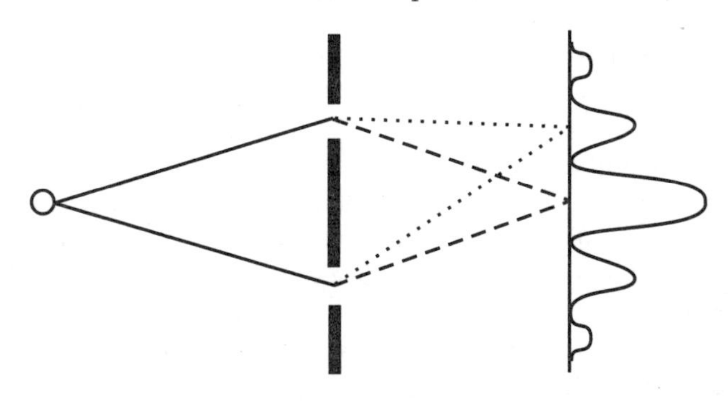

Fig. 2.3 Double-slit apparatus, or Young's interferometer.

the Mach–Zehnder apparatus correspond to the peaks of maximal intensity in the Young apparatus, and (ii) the differences in length over which all of the light is detected at *TT or RR* in the Mach–Zehnder apparatus correspond to the troughs of zero intensity in the Young apparatus.

Some of us see Young interferometry every day. If we happen, for example, to look at a streetlamp from far enough away through the fabric of a curtain of which the fibres have an appropriate spacing, the light from the streetlamp is distributed in a series of small points, the peaks of maximal intensity – the fibres form the slits and the screen is our eyes.

Thus, interference of paths is the fingerprint of wave behaviour; whence we understand now the name *wave mechanics* for the mechanics of atoms. Still, recall that atoms, and quantum particles in general, are not waves: as stressed already several times, particles don't split in halves at beam-splitters as waves do. In other words, as I have presented things in Chapter 1, starting with the notion of the particle, the astonishing result was obtained for the interferometers of **Figs 1.3** and **1.4**; had I presented the interferometers first, any physicist would have recognized a wave, but then the astonishing results would have been the individual detections. In some circumstances, quantum particles behave very much *like* corpuscles, and in other circumstances they behave *like* waves; but they are neither, as will become clear in Part 2 of this book.

2.2.3 Why quantum?

We now have an idea of the origin of the notion of the particle and of the invention of interferometry. But quantum physics was not discovered because a brilliant or clumsy physicist introduced, through inspiration or accident, some particles into an interferometer. The history is much more complex, as we are going to see. At the same time, we will discuss the origins of the adjective *quantum* that is applied as a qualifier designating this new physics.

The first indication, vague and at the time incomprehensible, appeared in the year 1900 in the works of Max Planck on the *blackbody*. The blackbody is an ideal object – a closed box that does not emit *any* radiation. A covered saucepan, for example, is not a blackbody. While it is true that no visible light escapes from it (we cannot see what is contained in the saucepan), some exchanges in heat are, fortunately, possible. The blackbody does not allow the passage of heat, visible light, nor X-rays ... In reality, we know of some physical objects that block a *wide range* of radiations, and therefore the model for the blackbody is a very good approximation of one. Now if we pierce a small hole in one of these blackbodies, the radiation that is emitted can be measured, and then studied with the aid of thermodynamic notions such as temperature, energy, etc. The theoretical description of this radiation was precisely the domain in which Planck worked at the beginning of the twentieth century. We see that it deals with waves that are propagated (radiation), it does not explicitly make mention of interference (although this does, in fact, play a role), and nothing at all of particles.

Planck noticed that he could explain certain observations that until then had defied any description, by introducing a peculiar *hypothesis*: the energy of a wave of a given frequency ν can only take certain values, namely the multiple of a certain minimal energy $E = \mathrm{h}\nu$, h being a constant. The hypothesis is peculiar because we are not accustomed to associating energy and frequency. In everyday life, the intensity of the light emitted by a bulb (energy) is independent of the fact that the light might be green, red or blue (frequency).

Planck baptized the hν minimal energy *quantum energy*, a Latin-sounding word evoking the idea of an elementary quantity. In one sense, Planck's hypothesis amounts to suggesting a corpuscular behaviour for the electromagnetic wave. This conclusion was explicitly drawn by Einstein during a lecture on the photoelectric effect that he delivered is Salzburg in 1909. The 'particle of light' (later to be called *photon*) re-entered physics one century after Young had dismissed it.

2.3 Continuation of the programme

We have reviewed the essential traits of the interference of a single particle as a physical phenomenon. We are going to pursue this same theme during the next three chapters, before moving on to describe the interferences of two particles. In addition to being useful for familiarizing ourselves with these concepts, the next pages will touch on some important questions: If all matter is composed of these quantum particles, why do we not see interference phenomena with cars, tennis balls, human beings? Why is it that physicists, and in particular the founders of quantum physics, have not succeeded in agreeing on an explanation of these phenomena? Is quantum interference a matter for laboratory life or is it open to application?

In the late afternoon, the setting sun is giving colour to the grey molasse of the old town of Fribourg. Here and there a window gleams, sending back part of the light of the sun into the eyes of strollers – blinded by the clarity of light.

·3·

Dimensions and boundaries

3.1 On real experiments

I am of the opinion that experimental physics is not given the credit it deserves in our society. Applied physics, ultimately industrial physics, is conveniently praised because it directly affects our lives. As for the high repute of theoretical physics, just stop and list five renowned physicists on the back of an envelope: I predict that the five names you have written down are those of five theoreticians (you can easily verify it on the internet). Experimentalists are those people who work in academic laboratories, confirming theories that they are (wrongly) supposed not to have understood, and making claims on possible applications that seldom attract the interest of industry. Of course, any physicist knows the truth: without the genius and the hard work of the experimentalists, there would simply be no physics at all. The laser is a striking example[20]: the basic equations had been written down by Einstein himself in 1917, but it required some brilliant experimentalists to make it real – against the scepticism of many colleagues, and without having any particular application in mind[21] . . . In this and the following chapters, while discussing, as promised, some deeper questions raised by quantum physics, I hope to also provide a convenient description of what a real experiment looks like – still, leaving aside the most thankless tasks: purchase of materials, days and nights spent making adjustments, error calculations . . .

For our first visit to the laboratory, we leave the Sarine for the Danube, the medieval charm of Fribourg for the imperial grandeur of Vienna. We are going down a side street with the evocative name of Boltzmanngasse, a reminder that during the first decades of the twentieth century, Vienna was one of the great capitals of Europe and this includes the realm of science. It has been said that Vienna gives the impression, even more so today than in the time of the Hapsburgs, of a theatrical setting[22]. In Boltzmanngasse, the sets do not merit a tourism detour, but the theatre is full of activity. We are going to attend two original shows created right here.

3.2 Neutron interferometry

3.2.1 One particle at a time

The experiments described in Chapter 1 have brought us to the surprising but inevitable conclusion that *each* quantum particle explores all of the indistinguishable paths. If that were not the case, it would be impossible to influence all of the particles by changing the length of only one path. But, in order that a real experiment allows us to arrive at this conclusion, it is necessary to ensure that there is *at most one particle at a time* in the interferometer. Indeed, if there were several particles at once in the apparatus, interferences may turn out to be an unwanted effect of the collisions between several particles.

This condition, having only one particle at a time in the interferometer, which is so essential to the interpretation of the results, is far from being easy to meet. We already know that in optics, the condition of having a single photon in the interferometer was met for the first time in around 1985 by Aspect and his co-workers in the experiment suggested by Feynman. It is generally admitted that the first series of interference experiments, which really showed single-particle effects, are those using neutrons performed by Helmut Rauch's group. These experiments started in 1974 – three quarters of a century after the intuition of Planck, fifty or so years after the work of Schrödinger, Heisenberg and Jordan who had set the basis of the formalism of the new theory, and as many years after the first experiment on electron interference performed by Davisson and Germer at the Bell Labs in New York. Until 1974, the hypothesis of collective effects to explain interference was acceptable, although not orthodox – after this date, it had to be dismissed as falsified by the experiments we are going to describe[23].

3.2.2 Source and interferometer

As mentioned, the particle studied in the Rauch experiments is the *neutron*, one of the components of the atomic nucleus. We begin by briefly describing the instruments that we see in the laboratory. The most cumbersome piece of equipment is the *source* of the neutrons. There are not many ways in which neutrons can be emitted – it is necessary to break the nucleus of certain atoms, a nuclear reaction that bears the name *fission*. Therefore, a source of neutrons is a *nuclear reactor*. The reactor does not have to be powerful – we only want one neutron at a time in the interferometer, therefore it is better that the reactor produces few neutrons. In the laboratories of Vienna, there was one of these small reactors, but by 1975 Rauch and his colleagues will be doing their experiments in a better equipped laboratory, the Laue-Langevin Institute in Grenoble.

The *interferometer* (**Fig. 3.1**) is quite small in comparison with the reactor, its dimensions measured in centimetres. But putting aside the fact, to which we will return, that even centimetres are enormous dimensions in relation to the size of a neutron, this little black object is a technical feat in itself, because it is a *monocrystal*. A small

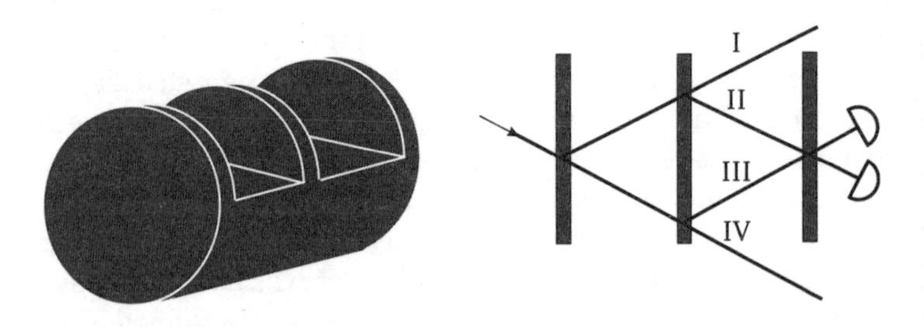

Fig. 3.1 Mach–Zehnder type interferometer for the Rauch experiment: geometry of the cut silicon monocrystal and scheme of possible paths.

digression is useful in order to explain this term and understand where the feat is.

In solids, the atoms are arranged in ordered structures. For certain crystals, this ordered structure can be seen with the naked eye, the better-known examples being prisms with hexagonal cross-sections formed by quartz, or tiny cubes of cooking salt. Each of these quartz prisms or cubes of salt is a monocrystal – its atoms are ordered according to the ideal structure. Large crystals consist of several of these prisms, which are formed in different directions because of growth defects. Therefore, in general, a monocrystal is one whole part of a crystal in which the atoms are arranged according to the ideal structure, without defects, and a crystal is formed by several monocrystals separated by growth-defect regions. The interferometer for neutrons must be cut into a silicon monocrystal for a reason that will become apparent soon. But nature does not produce silicon monocrystals that big, so a careful growth procedure has to be used to grow the monocrystal in the laboratory – the technique of crystal growth is something very 'technical', but becomes a necessary pre-requisite for 'fundamental' studies.

A relatively fine strip of silicon (half a centimetre) acts as the beam splitter for the neutrons. Note that the two perfect mirrors, which appear in the schematic apparatus in Chapter 1, are replaced here by two beam splitters – in fact by only one beam splitter, placed across both paths. For the results of the experiment, the presence of the beam splitter in place of the mirrors implies simply that half of the particles are going to take paths I and IV, and do not participate in the interference, the indistinguishable paths being II and III.

It remains to understand why it is necessary that the atoms are arranged in a regular way throughout the interferometer, why the whole interferometer must be a monocrystal. This still calls on the notion of interference, but at a level that we have not yet encountered. The situation is outlined in **Fig. 3.2.** The beam of neutrons, or the space in which we can find a neutron now and again, extends over a certain area (the grey strip in the figure). Consequently, some interference phenomena can be produced between the different paths that we can illustrate in this beam. In the figure,

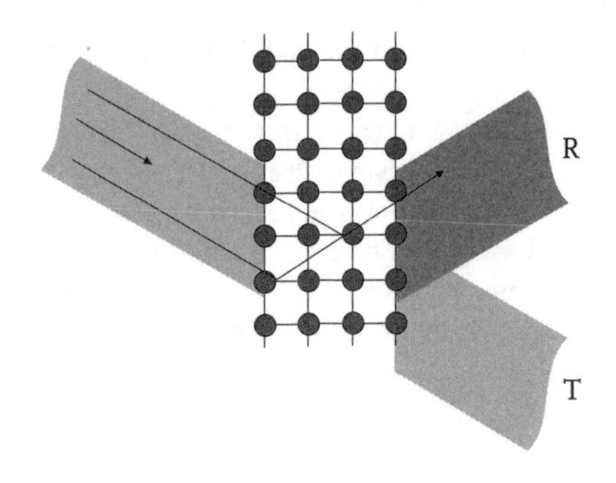

Fig. 3.2 The interference principle on each atomic plane of the crystal (Bragg's reflection).

we have drawn two possible paths in the beam that, after reflection by a beam splitter, merge with one another and therefore become indistinguishable. Having noted that, we must apply the usual reasoning to the interference (see Young's slits): depending on the difference in the paths taken, the interference could be destructive, in that we might not find any neutrons at the output in that direction. We see in the diagram that, the distance between the atomic planes being determined by nature, the difference between the paths depends on the *angle* at which the beam of neutrons arrives at the atomic plane. In order to build an interferometer, it is necessary to be able to control the angle of impact of each part of the beam with the atomic planes at each beam splitter. It is therefore essential that the atomic planes have the same orientation in the three beam splitters, and this amounts to requiring that the interferometer be cut into a monocrystal.

Having described in some detail the technical difficulties, we can go back over the goals of the experiment that concerns us here, because there is also something new to learn in the physics. We have just seen that the geometry of the interferometer is extremely rigid – moving a single atomic plane in the beam splitter is out of the question! And between two beam splitters, the neutron travels through the air without any guidance. But in Chapter 1 we have seen that in order to conclusively observe interference *it is necessary to alter one of the paths* (the experiment of **Fig. 1.4**). How are we going to set about doing it? The reader will benefit greatly if, before going onto the next paragraph, they devote some time to looking for the answer themselves – physicists have received sufficient information in their training to get there, for others the challenge is more difficult but it is not a mission impossible.

3.2.3 The differences between the paths

Rauch's apparatus allows us to perfect our understanding of interference. As we have seen, we talk about interference when the results of detection depend on the difference between indistinguishable paths. In the introductory account, the difference between the paths was a *difference in length* – one path is longer than the other. The difference in the length is also the origin of the interference in the double-slit apparatus. The key to the success as regards neutron interferometry lies in the fact that it is not necessary to change the length of a path – it is possible to keep the whole interferometer intact and change *something else* on one of the two paths. The important thing is that the two paths are *different for the particle.* There you have the answer to the question posed above (it was not too difficult to find it, was it?). Now we see how this difference is obtained for the neutrons in the Rauch apparatus.

The new fact that it is necessary to take into account is that neutrons possess a property that we call *spin.* 'To have a spin' means the following: in addition to having a position and a state of motion, each neutron possesses a property linked to a direction in space. This is shown schematically in **Fig. 3.3**, where the neutron is illustrated as a small ball pierced by an arrow – the arrow represents the spin, the information on a spatial direction that the neutron carries. Naturally, the image of the arrow is inadequate for representing the spin, just as the image of the ball does not represent the other properties of the neutron very well – we already know that

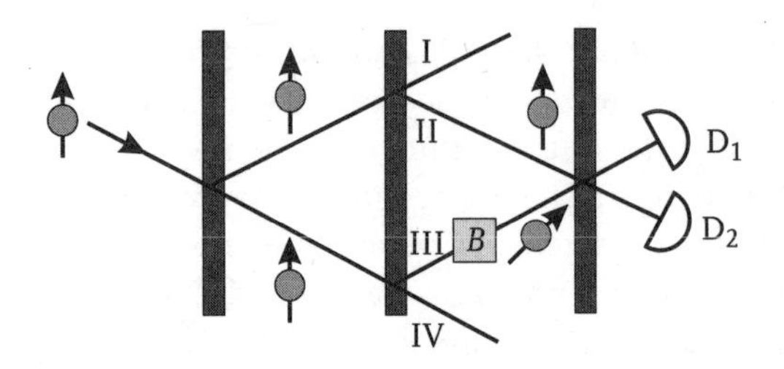

Fig. 3.3 The Rauch interferometer with a magnetic field B that modifies the direction of the spin on path III.

this very special 'ball' explores several paths. For example, next we are going to use the fact that *two 'arrows' (directions of spin) are distinguishable with certainty only when they are opposite.* And there we have a property of these 'spin arrows' that is not viable for road direction indicators – happily, for that matter! It is extremely difficult for a physicist to say what things *are*, but on the other hand, with the aid of the arrow, I can at least explain *how things work*.

In order to introduce a difference between paths II and III, the experimentalists of Rauch's group *turn the spin* on path III. In practice, this is done by introducing a magnetic field B into the path. After this rotation, the direction of the arrow is generally different on path II and on path III: the two paths become different for the particle.

The physicist now wishes to quantify this difference. In Chapter 1, it was intuitive to choose the difference of lengths as a measure for the difference in the paths. Here, it seems natural to measure the difference in the paths by the angle between the arrows. This intuition is not bad, but it turns out that it is *half* of the angle between the arrows that measures the difference! It sounds strange, but this is the prediction of quantum theory – a prediction that we can understand well enough with what we already know from Chapter 1. In fact:

- In the case where we change nothing on path III, the two paths are identical, therefore all of the particles will arrive at D_2.

- Let's suppose now that we turn the arrow on path III by 180°. In this case, the difference between the two arrows is as large as possible – the two arrows are opposites. But we have said that in this case (and only in this case) the two paths become distinguishable – by measuring the spin, one can know with certainly by which path the particle arrived. Therefore the particles must behave classically – half are detected at D_1, the other half at D_2.
- Finally, by doubling the difference between the paths in comparison with the case above, we arrive at a situation where all of the particles are detected at D_1, the opposite to the initial situation. This situation is therefore reached when the arrow is turned 360°, or, when it does a complete turn! The arrow must do two complete turns in order to return to the initial situation, with all of the particles detected at D_2.

This behaviour has surprised physicists, who have given it the name 4π *spinor symmetry* because in scholarly language 4π signifies 'two turns', one turn corresponding to 360°, that is to say 2π in radians[24]. To predict this behaviour of the spin, we have only used our general knowledge of interferometers and the condition of distinguishability of these 'arrows'.

3.2.4 The dimensions of Rauch's interferometer

The Rauch experiments have shown that quantum physics applies to experiments performed on single systems[25]. Before moving on to something else, I promised to discuss in more detail some matters of size. Namely, since a particle explores several paths at a time, it is legitimate to wonder about the size of the particle in relation to the size of the apparatus. In itself, the question is not very well set out, as, in principle, the paths of the interferometer can be of arbitrary length, it is only their difference that can play a role. The figures for Rauch's apparatus are nevertheless impressive. Here they are.

The size of a neutron in the nucleus of an atom is measured at 10^{-15} m, a length of which we have no sensory perception – it is ten to one hundred thousand times smaller than an atom, the size of an atom being measured in turn in billionths of metres, that is in millionths of millimetres! Let's remind ourselves that the human eye can distinguish, under the best conditions, two points separated by a tenth of a millimetre (10^{-4} m) and that visible light (that is, the best optic microscope) can only probe distances in the order of a thousandth of a millimetre (10^{-6} m). The distance between the two paths is some 2 cm, the area of the space in which the magnetic field is present around path III is in the vicinity of 2×2 mm^2. We can therefore acknowledge that the influence of this field at path II is negligible, and that is what is desired. Finally, the rate of emission from the source is such that the distance between two neutrons emitted successively would correspond to 300 m. Since the interferometer measures less than 10 cm, we can safely assume that there is only one neutron at a time in the apparatus.

These numbers mean then that if the neutron were the size of a coin, the distance between the two paths in the Rauch experiment would be comparable to the distance between the Earth and the Sun. If the magnetic field, which serves to turn the spin on path III, were centred on the Sun, the influence of the field would already be negligible at the orbit of Mercury. And yet, the proportions being thus, by modifying only one of the two paths we can modify the behaviour of all of the neutrons! Obviously, quantum particles do not behave like coins or cars . . . or like footballs, even when they have that shape, as we are going to see in the next part of this chapter.

3.3 Interference with large molecules

3.3.1 The career of a student of Rauch

If we cast our eye over the list of Rauch's collaborators twenty years later, one name leaps out at us – that of Anton Zeilinger. The quantum optics group that Zeilinger set up in Innsbruck during the latter half of the 1980s is credited with some remarkable experiments, to which we will return later in this book. In 1998, Zeilinger was named professor at Vienna and he moved his entire group from the heart of the Alps to the capital. Just before leaving Innsbruck, he had engaged a young collaborator, Markus Arndt, to whom he proposed embarking on a new experimental project – the observation of quantum interference for *large molecules.*

3.3.2 Researching the boundaries

We know by now that quantum particles (atoms, neutrons, electrons, etc.) produce some interference effects. We also know that real-life objects (cars, coins, etc.) do not produce such effects. And we have seen in Chapter 2 how the difference is closely related to our customary way of describing properties as a set, which fails for quantum systems. Since the beginnings of quantum physics, the transition between the 'classical' world (the everyday) and the 'quantum' world has been a subject of debate. Among the questions that one can formulate are the following: If we indeed suppose that quantum systems are the building blocks of everything, how is it that the classical world arises on such a large scale? Are we human beings, accustomed to the classical world since Adam and Eve, ever to 'grasp' the quantum behaviour? These concerns are not settled yet. I will focus on the first one.

To that first question, many physicists answer that the transition from the quantum to the classical should be a *factual* transition, not an essential one. There is no law of physics preventing cars from interfering; only that such an effect is extremely difficult to show in practice. In other words, according to this vision of reality, there is no clear boundary between the quantum and the classical world, everything being fundamentally quantum, the only problem being the ability to probe the 'quantumness' of everyday objects. On the one hand, as no known law or principle of physics is contradicted, this vision goes unchallenged; on the other hand, some physicists (like me) find the extrapolation slightly dramatic[26].

In any case, everyone agrees that we are touching on one of the most fundamental open questions in quantum physics. Rivers of ink have been spilled proposing purely conceptual solutions to these problems. We must therefore rejoice in the experimental

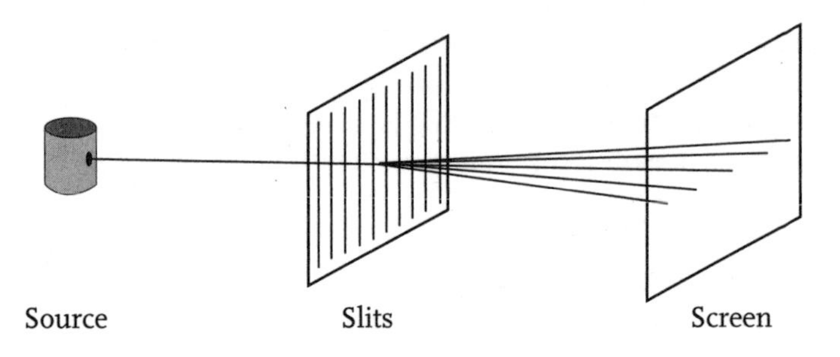

Fig. 3.4 Diagram of the experimental apparatus that demonstrates Young interference for C_{60} molecules.

works that offer to provide some answers, as partial as they might be. It is one of the most recent of these experimental works that I would like to examine in closing this chapter[27].

3.3.3 The experiment performed in Vienna

In the experimental research into the transition between the classical world and the quantum world, Zeilinger and Arndt have taken an important step. They have shown that some *large molecules produce interference effects* (**Fig. 3.4**). The molecules in question are collections of sixty carbon atoms and their symbol is C_{60}. In these molecules, discovered in 1985, the atoms are arranged according to a particular symmetry, that of a *football* – indeed a traditional football, formed with hexagons and pentagons stitched together, having exactly sixty vertices, that is sixty points where three lines meet. Sixty carbon atoms arrange themselves according to the same structure in order to form C_{60} molecules.

Such beautiful molecules deserved a name other than that of their chemical composition. At the moment of baptism, the scientists were reminded of the work of Richard Buckminster Fuller, an American architect who had designed and built numerous glass domes whose supporting structure has the symmetry that we are talking about[28]. It is therefore in memory of Buckminster Fuller that the C_{60} molecules are not called *footballenes*, but *fullerenes*, and sometimes *buckyballs*.

As far as their size is concerned, fullerenes are clearly closer to atoms than cars or footballs and in this sense we are not surprised to see them display quantum behaviour. Nevertheless, the criterion for the observation of quantum behaviour is not the small size of the physical object, but the possibility of creating a situation of indistinguishability. Viewed from this angle, we better understand that the interference of large molecules constitutes an important result. In effect, the bigger the molecule, the more chance there is that some or other of its constituent parts will interact with the environment, and if on one of the possible paths a non-controlled interaction takes place, the interference is quickly lost. Now, a molecule with sixty carbon atoms means a system of sixty nuclei – which for carbon means 360 protons and as many neutrons – and 360 electrons. In total, 1080 'elementary' quantum particles (I am disregarding the fact that protons and neutrons are in turn composed of three quarks each, because that final composition has some characteristics that merit a separate discussion).

The experimental observation of fullerene interference[29] should not be considered as just a further verification of the quantum behaviour of matter, but rather as a real *discovery*[30] – we are far from demonstrating the interference of cars, but these are already large objects: it was not evident *a priori* that such a large collection of quantum particles would itself exhibit collective quantum behaviour[31]. The way is open to investigating much bigger molecules, such as insulin and other 'biological' molecules. The rush for size has just begun.

3.3.4 Quantum football

On the internet page of Zeilinger's group, the C_{60} project is introduced by an animation. We see a football player kick a ball that passes both to the right and the left of the astonished goalkeeper (dressed as a physicist for the occasion), to end up in the goal. Nobody thinks that this animation illustrates the football of the future – the Maradonas (or the Herzogs, to do justice to local talents) of the coming years will have to be content with going around walls either to the right or the left. The little animation is meant to be paradoxical and in fact it clearly illustrates the difference between everyday objects and quantum objects. It invites us to ask ourselves the question 'Where is the boundary, and is there really one?' The question remains very vague, but after a century of quantum physics we are finally in a position to do the experiments to glimpse a solution.

·4·
Authority contradicted

4.1 The Heisenberg mechanism

4.1.1 Constance, 1998

Physicists have their moments of weariness, enthusiasm, apprehension, relief, just like everyone in their work. Weariness, that is when something does not work – when the apparatus does not want to function or when the last bit of the theorem resists every attempt at demonstration. Enthusiasm, that is surely when we finally find an error or we overcome an obstacle. The apprehension phase generally begins when we must present our latest results to the scientific community, by writing an article or participating at a conference. That is the moment of truth, but also sometimes the moment of low blows, real or imagined.

In the case of Gerhard Rempe and his colleagues at the University of Constance, a certain amount of apprehension is entirely justified. In the month of March 1998, they are waiting for their latest article to be reviewed by the referees chosen by the prestigious scientific journal *Nature*. They are convinced that it is a quality work. But they also know that their results challenge the explanation of Heisenberg's celebrated 'uncertainty principle' proposed by Werner Heisenberg himself[32], just as it was (with variations) by Einstein and many others. We will talk about the 'Heisenberg mechanism' in order to clarify the ideas.

4.1.2 Beyond principles, a mechanism?

What does this all mean exactly? We have seen that if we determine by which path (or for the Young experiment, through which slit) the

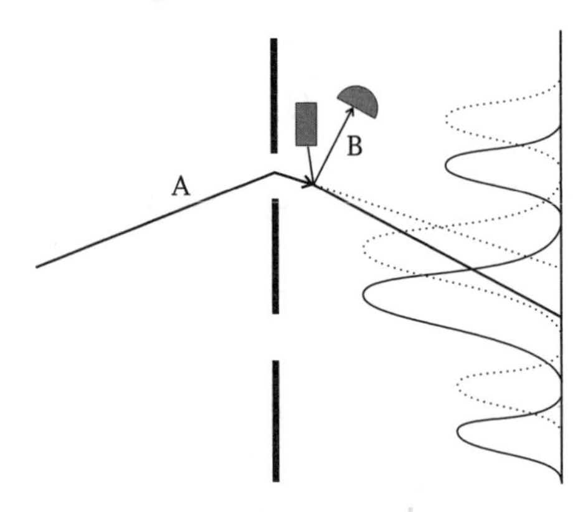

Fig. 4.1 Illustration of the Heisenberg mechanism for the Young interferometer. The probe particle (B) is supposed to be much lighter than the particle to be measured (A).

particle travels, the interference disappears. This is consistent with the indistinguishability principle – in making a measurement, we remove all possibilities but one, therefore the particle behaves in a classical manner. There we have a *description* that leads to the right prediction. But could we not find an *explanation*? More precisely, could we describe this phenomenon not with the aid of a principle, but with a *mechanism*?

Heisenberg's argument, illustrated in **Fig. 4.1**, aims to do that. In order to measure the position of particle A after its passage through the slits, it is necessary that it interact with (at least) one other particle, particle B, which we detect after the collision. Now, in a collision (consider the collision of two billiard balls) the trajectory of *both objects* is modified. Likewise, says Heisenberg, the trajectory of particle A that passes through the slits is modified by the collision with particle B that acts as a probe. Since we cannot control the details of the collision, particle A is deflected unpredictably, and the interference disappears. In fact, the interference has not really disappeared – if we could control the collision, and select only the particles that have undergone a very precise collision, according to Heisenberg the interference would appear again. But since we cannot sort events that way, all that we record is the superposition of several interference patterns displaced among one another: this is a smooth curve.

Nobody can deny that the argument is ingenious. In fact, the idea that the introduction of a random element (here, the collision is impossible to control with any precision) obliterates the interference is correct. For example, the interference pattern for C_{60} molecules observed by Zeilinger's group in their first experiment was not very pretty because the molecules had not been selected according to their speed, and speed is one of the parameters that determine the position of the peaks of maximal intensity in the interference fringe. Once the distribution of speed was narrowed through a filter, the interference fringe improved markedly – the peaks were better defined. The question then is not of knowing if a non-controlled distribution obliterates the interference (it does), but if *all* disappearance of interference can be attributed to a similar mechanism that we cannot control.

Given what we already know, we can legitimately harbour some suspicions. In particular, in Heisenberg's argument we suppose that particle A follows a well-defined trajectory, a trajectory that measurement only would reveal, before modifying it. We have already seen that the existence of a trajectory is not at all obvious. In particular, the apparatus in **Fig. 1.4** showed that it is enough to change the length of a path to modify the behaviour of all of the particles; and in Chapter 2, we argued that the 'which path' question cannot be answered in an interferometer.

That said, let's examine Heisenberg's argument more closely anyway: how does it differ from the *indistinguishability principle* that we have discussed? According to our principle, interference disappears as soon as the two paths are distinguishable, whatever the cause. Heisenberg proposes *one very precise cause* – the crudeness of our measurements, which prevents us from having access to all of the significant parameters. If we were capable of making more precise measurements, we would not lose the interference. Unhappily, suggests the German physicist, we are not capable. The challenge is thrown down: will we be capable of it one day, or is nature (of which, let's not forget, we are a part, as are our measurement devices) designed to keep this issue from being resolved?

4.2 The Heisenberg mechanism in the laboratory

In order to put the Heisenberg mechanism to the test of an experiment, it is necessary to show that we can introduce distinguishability between two paths without modifying the trajectory (position and speed of the particle over time) in any significant way. Once this condition is met, it will suffice to look at the screen or the detectors: if the interference remains, Heisenberg was right; if it disappears, he was wrong.

The idea seems simple and clear cut, but the trap is hidden in the words 'in any significant way'. Some modification of the trajectory of a particle is unavoidable from the moment we try to interact with it. The problem that presented itself to the physicists at Constance was this: to introduce a clear distinguishability between the paths, in such a way that the modification of the trajectory, which unavoidably ensues, is nevertheless too small to explain the complete disappearance of the interference fringe.

4.2.1 The interferometry of atoms

The apparatus produced at Constance is a Mach–Zehnder interferometer, like that of Rauch. The particles used are atoms, rubidium atoms, to be precise. Atoms are quantum objects consisting of a *nucleus*, heavy and carrying a positive electric charge, and *electrons*, much lighter, negatively charged particles. We know this thanks to the well-known symbol in which the atom is represented like a small solar system with a few planet-electrons orbiting around a sun-nucleus. Again, as with the arrows we used to represent spin: the electrons and the nucleus themselves being quantum particles, this symbol is only a pale imitation of what an atom really is; but it is a useful picture to keep in mind.

From this structure of the atom some consequences ensue that are significant for the goal that interests us. On the one hand, the trajectory of the atom is essentially determined by the motion of the nucleus (if we can continue with the planetary analogy, we see clearly that the orbit of the enormous Jupiter around the Sun is only slightly affected by the presence of the satellites that gravitate around the planet). Consequently, in the experiment that we want to design, the beam splitters can be devised to act on the nucleus; and the electrons will follow the motion. On the other hand, it is relatively easy to modify the physical state of an electron, particularly that of 'external' electrons (those that are furthest from the nucleus). Therein lies the solution: it is by modifying the state of an electron on the path, or to be precise, its energy, that we will be able to introduce distinguishability without influencing the motion of the nucleus.

Figure 4.2 illustrates the results of the experiment. The part on the left of the figure represents the initial apparatus, with the number of particles detected behind each output. We observe an interference fringe characterized by the fact that the peaks of

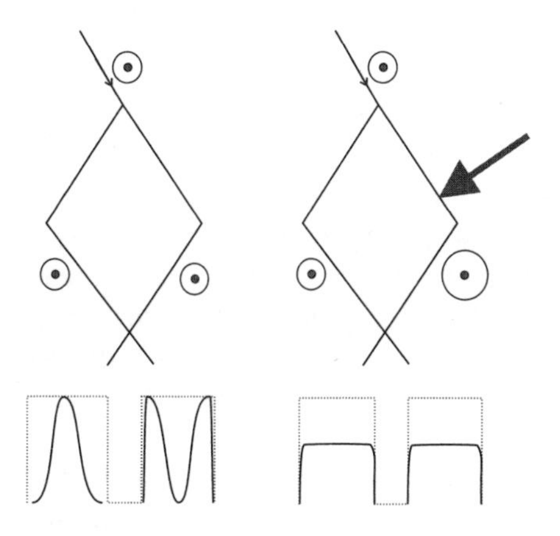

Fig. 4.2 Diagram of the Constance experiment: distinguishability is introduced by modifying the properties of the electrons on one of the two paths.

intensity are complementary on either side, that is, a peak to the right corresponds to a trough to the left and vice versa. The part on the right of the figure represents the modified apparatus – on one of the two paths, the energy of an external electron has been modified. Now, by measuring the energy of the electron we are able to learn which path it took. It is not necessary to add an instrument that effectively measures this energy to the apparatus – the important thing is that we have introduced distinguishability. The information encoded in the atom is sufficient for us to be able *in principle* to discern the two paths. As we see in the figure, *the interference disappears*. Thus, the mechanism proposed by Heisenberg does not explain the disappearance of interference completely. We must be content (for the moment at least) with the indistinguishability principle.

4.2.2 The significance of the Constance results

The Constance experiment is a beautiful experimental achievement – we have not spent words about the technicalities of atom interferometry as we did for neutrons in Chapter 3, so let me just say that the requirements are just as stringent, if not more so. From the conceptual point of view, I think that the importance of this experiment is mainly didactical. Recall once more the exchange between Feynman and Aspect: the experiment proposed by Feynman was not aimed at demonstrating something new, but at *directly* demonstrating something that had already been assumed on the basis of much indirect evidence. The situation is similar here – indeed, at the very beginning of this chapter, we had already guessed that Heisenberg's mechanism was bound to fail, based on what we knew of quantum phenomena. A direct demonstration is nevertheless useful.

Historical reasons add weight to this argument. Heisenberg's mechanism is precisely that, a mechanism: therein lies its appeal. From the very beginning of quantum mechanics, it was used by many people to illustrate what happens in the quantum world, and is expounded in the majority of treatises on quantum physics and university courses[33]. Notwithstanding some words of warning, which appear in most good textbooks, many physicists did not notice the limited value of the Heisenberg mechanism. The removal of this frequent misunderstanding fully justifies the experiment performed in Constance.

4.3 Complementarity and uncertainty

We are all a bit disappointed by the failure of the Heisenberg mechanism. When exposing it, we had a sense of understanding, of feeling a bit better about what was happening – had it been correct, it would have provided a mechanism, an explanation fitting well enough with our images, with everyday language. I would like to end this chapter with a brief mention of these language problems.

Since the advent of quantum physics, physicists have carried with them the heavy baggage of words that are supposed to convey the concepts of the new science. The words 'complementarity' and 'uncertainty' make up part of this baggage. Their survival in the world of physics is not assured.

The word *complementarity* was forged by Niels Bohr. The concept that it conveys is closely linked with the indistinguishability principle that has been discussed in this book. In order to clarify the idea, let's go back to the apparatus of **Fig. 1.3**. Our description was this: if we do not know by which path the particles travel in the interferometer (two indistinguishable paths), all of the particles take a certain output (interference); if we detect the particles in the interferometer (distinguishability), the output will be random. Bohr would say instead that the *path* and the *output* are two pieces of *complementary information* – we cannot arrange it so that all of the particles take the same path *and* the same output. At the risk of missing something very profound, we will remember that Bohr's complementarity principle says the same thing as the indistinguishability principle, from a different angle. The future will tell us if one of these concepts will disappear to the benefit of the other, if they are destined to survive together, or if both will be erased by new, more precise notions.

The word *uncertainty*, as far as it goes, is unfortunate because in physics we already use it to describe the imprecision of measurements. If a length is measured with a ruler graduated in millimetres, the value that one reads is affected by an uncertainty of (give or take) a millimetre. In other words, with that ruler, one cannot discriminate two lengths that differ by less than a millimetre. The Heisenberg mechanism is an attempt to restore a principle of uncertainty in measurement to quantum physics, an attempt, in other words, to base the wealth of phenomena that we encounter in quantum physics on our technical limitations (essential or accidental). From experiments like that of Constance we learn that the principle of quantum physics is not a principle of uncertainty in that sense, rather a *principle of indetermination* – as precise as our measurements are, we will not be able to determine two pieces of information that are complementary in the Bohr sense. Quite the opposite, we have always worked under the assumption of perfect detectors – imperfect detectors could introduce so many counting errors that the quantum interference would be masked. In summary, the concept of uncertainty is ambiguous in physics; and, if we want to retain the traditional sense of 'limitation in the precision of measurement', then this concept is not adequate to describe quantum behaviour.

Complementarity, indistinguishability, uncertainty, indetermination... the physicist sometimes plays the wordsmith, the poet who seeks the best imagery. For now, Rempe and his colleagues have another concern. They remember the controversy in *Nature* that gave rise to several articles and some experimental works – theirs will be the final word, but they do not know it yet. They remember the constructive discussions that they had, but also a few physicists who had found the idea of challenging the Heisenberg mechanism simply scandalous. Let's hope one of them is not a referee for our article, they say[34].

·5·
A nice idea

5.1 Bangalore, December 1984

For the last chapter devoted to the interference of a particle, we go back a few years in time, to 1984. If until now everything has taken place in the familiar setting of old Europe (Fribourg, Vienna, Grenoble, Constance), this time the change of scene will be much greater – this year, the *IEEE International Conference on Computers, Systems and Signal Processing*, one of the most important gatherings of computer scientists and communication theoreticians, is taking place in the city of Bangalore, at the heart of the immense triangle that is India.

However, all conference rooms look the same. Gilles Brassard, computer scientist from the University of Montreal, is not too disoriented when he is preparing himself to present a project done in collaboration with Charles Bennett, trained as a chemist, currently a theoretician at IBM in New York. The title of the lecture arouses the curiosity of those who glance absent-mindedly through the conference programme: Brassard is going to speak on 'quantum cryptography'. For the majority of people present in the room, the second word does not hold any surprise; the first, on the other hand, is shrouded in an aura of mystery. Quantum physics, is that not that bizarre theory on which physicists themselves ceaselessly debate? Gilles Brassard begins his lecture by 'recalling' the basic principles of quantum physics – in reality, he knows very well that it is not a reminder, but a novelty, and that the success of his lecture is going to depend largely on the way in which his introduction is understood. Whilst Brassard displays his pedagogic abilities, we will leave the room because the reader of this book is beginning to have an idea of what quantum physics is, but *a priori* does not necessarily know what cryptography is. In the corridor, with one eye on the progress of the lecture, I am going to endeavour to fill in the gap in my reader's knowledge.

5.2 Cryptography

5.2.1 The birth of a science

The art of sending secret messages, in war, in love or simply in fun, has been practiced since the dawn of time, but it was not until the twentieth century that the underlying rules of this art were systematically studied. This new science was named cryptography. *Cryptography* (from the Greek, 'secret writing') is therefore a scientific approach to one of the oldest problems in the world – how to send a coded message, which only the people authorized to do so could decipher. History has kept track of countless cryptographic tricks. These days, there are two main classes of cryptographic methods – public and secret key protocols. The first type are used primarily for authentication (electronic signatures, for example) or for the transmission of messages to numerous parties, the second type for the transmission of messages between fewer parties – typically two, Alice and Bob. Let's leave the public key protocols to one side and describe a secret key protocol, the one that inspired Bennett and Brassard.

5.2.2 The one-time pad, or Vernam code

The techniques for sending secret messages that we practiced in our childhood are elementary examples of secret key cryptographic protocols. It is a question of scrambling the message that we want to send according to rules of which only the authorized parties should be aware. In fact, however, it is not easy to construct an absolutely secure code. The common method consisting of substituting letters (replace A with C, B with T...) is not secure, because in every language the letters have a determined frequency. For example, in a reasonably long message written in English, the most frequent letter is theoretically 'e', and for a computer, it would not be difficult to decipher a coded message of this type. But in the class of secret key protocols, one exists that is absolutely secure. It was invented by Vernam in 1927, and bears the name *Vernam code*, or *one-time pad*. The principle is simple – it took some thinking!

The authorized parties Alice and Bob, and only them, must possess from the outset the same long list of *bits* – a *bit* is what we call a variable that can take only the values 0 or 1. This list is the secret *key*. Alice, who is sending the message, begins by writing the message in binary code. In order to clarify the idea, we will use ASCII code, in which we associate to each letter a list of eight bits. For example, the message JE T'AIME (French, *I love you*) composed of seven letters (we ignore the space and the apostrophe) becomes the following list of fifty-six bits:

J	E	T	A	I	M	E
01001010	01000101	01010100	01000001	01001001	01001101	01000101

Then we draw at random a list of fifty-six other bits – this list is the *key*, which must be known only by the parties Alice and Bob. Alice, who is sending the message, adds up the message and the key bit by bit, using the convention of binary addition $1+1=0$. The sum of the message and the key gives the *public text*. For example:

Message	01001010010001010101010001000001010010010100110101000101
Key	01010011010101101010011101010001011111010101110011100101
Public Text	00011001000100111111001100010000001101000001000110100000

If the key is indeed random, the public text does not contain *any information* about the message. In effect, a 0 in the public text can correspond to a 0 in the message (if the bit in the key is 0), or to a 1 in the message (if the bit in the key is 1). Therefore, we can shout the public message from the rooftops or put it up on the internet – nobody will understand what it means, unless they have the key.

So the Vernam code is unbreakable... provided only the authorized parties have the key! The problem of security therefore lies with the distribution of the key: how are Alice and Bob going to manage to construct a key that they, and only they, know? We are used to the spy movie method of 'exchange of briefcases' and the risks involved in this method. Alice and Bob certainly cannot use the telephone, an eavesdropper could have tapped the phone... could there at least be a way of knowing if an eavesdropper was listening? It is time to go back into the room – Brassard has finished presenting those elements of quantum physics of which he had a need, and is going to answer our last question.

5.3 Quantum key distribution

5.3.1 The principle

The knowledge that we have of quantum physics is enough to understand quantum cryptography. The principal idea is as follows: Alice sends Bob some quantum particles. We have seen that if we detect by which path a particle travels in an interferometer, the interference disappears. This idea that '(intermediate) measurement modifies the results' is what we use to get around the eavesdropper – if an eavesdropper makes a measurement on the channel linking Alice to Bob to try to get information, they are going to notice it, since their results will not be the results they expected. We take advantage, then, of the indistinguishability principle to detect for certain the presence of an eavesdropper who has tapped the line.

Once the idea is grasped, it is instructive to follow a protocol in its details. The protocol proposed by Bennett and Brassard in Bangalore is the example best suited to this task.

5.3.2 The apparatus and the protocol

By using the apparatus with which we have become familiar, the quantum key distribution invented by Bennett and Brassard is as

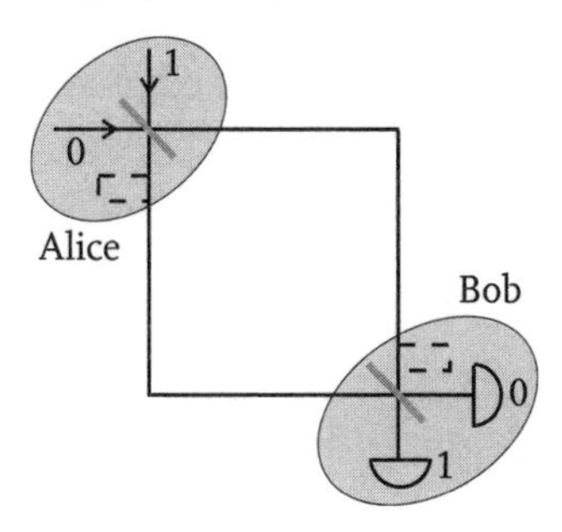

Fig. 5.1 Diagram of the principle of quantum cryptography: in dashes, the extensions that Alice and Bob can add.

follows. Alice and Bob find themselves at the two extremities of a Mach–Zehnder interferometer in which the two paths are of identical length[35]. Furthermore, each of the parties can choose to lengthen one of the two paths a little. Following on from the discussion in Chapter 1, the following situations could be produced (**Fig. 5.1**):

1. When two extensions are in place, or when no extensions are in place, the paths are strictly identical. Therefore if Alice sends a particle by input 0, Bob will receive it at output 0; and if Alice sends a particle by input 1, Bob will receive it at output 1.
2. When only one extension is in place, the paths are different. The length of the extensions are adapted so that, whatever input is chosen by Alice, Bob has a 50% chance of finding the particle at output 0, and 50% of finding it at output 1.

That is all for the physics. We can now explain the quantum key distribution protocol. This protocol comprises the three following steps:

Alice sends Bob some particles. For each particle, Alice chooses at random one of the two inputs, and also chooses at random whether or not to insert the extension. Bob chooses equally at random, and independently of Alice's choice, whether or not he is going to insert an extension. At the end of this step, each party has a list of bits, and for each bit they know whether or not they have inserted an extension. For example:

Alice

Particle #	Bit (input)	Extension
1	0	Yes
2	0	No
3	1	No
4	0	No
5	1	No
6	1	Yes

Bob

Particle #	Bit (output)	Extension
1	0	Yes
2	0	Yes
3	1	No
4	1	Yes
5	1	No
6	0	No

As we have said above, each time the content of the *Extension* column is identical, Alice and Bob have the same bit – when the content of the *Extension* column is different, in 50% of cases Alice's bit will be different from Bob's.

Public communication. Alice and Bob publicly communicate to each other (this communication is not secret, it can be heard by anybody) the contents of their respective *Extension* columns. They discard every case where the content of this column is not the same – in the preceding example, they discard bits 2, 4 and 6, and retain bits 1, 3 and 5. These bits are identical, and they are secret – the contents of the *Bit* column is never revealed.

Verification. Finally, Alice and Bob publicly reveal some of their secret bits (which are then no longer secret). If the value of these bits is always identical for both parties, they conclude that the transmission of particles has worked well – the other bits, those that were not revealed at this step, will form the key. In the above example, Alice and Bob can reveal bit 1 (value 0 for both) and bit 5 (value 1 for both). Since the value coincides for bits 1 and 5, they can be confident that the value of bit 3 is also the same for both: bit 3 can make up part of the secret key.

Having made the effort to understand this protocol, the reader is certainly asking themselves, why this game with extensions? Really, for the moment these extensions have only served to discard half of the bits, and so slow down distribution of the key! We are going to see now that the game of extensions allows Alice and Bob to detect the presence of a possible eavesdropper who may have bugged the quantum line (the interferometer).

5.3.3 The eavesdropper does not go unnoticed

Let's suppose now that between Alice and Bob there is an eavesdropper, Eve, who is going to attempt to discover the key. We will see that Eve can easily get a fair amount of information, but in doing so she will inevitably introduce some errors into Bob's results, which will be noticed. In other words, Eve can tap the quantum line, but Alice and Bob will be aware of it and stop distribution of the key.

Let's try to prove to ourselves the fact that Eve cannot go unnoticed, by considering one possible attack strategy, which is not the optimal strategy but is good enough to understand the principle. This strategy is described in **Fig. 5.2**: Eve cuts the lines and diverts the particles to her own interferometer. In this way, she can make the measurement, as Bob would have done. However, Eve's task does not stop there. Bob is waiting to receive a particle,

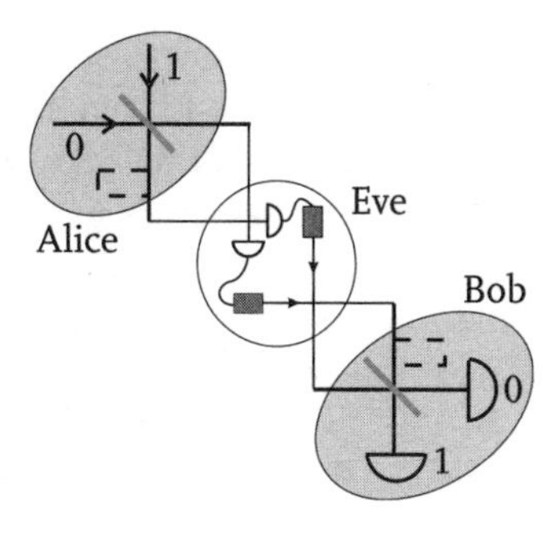

Fig. 5.2 A possible attack for the eavesdropper Eve.

the possibility that Eve is substituting herself for Bob has not been considered – to put it bluntly, no cryptographic strategy can stop Eve from killing Bob and passing herself off as him!

So what can Eve do? The best thing she can do, naturally, is to imitate Alice by creating a new particle and sending it to Bob through an interferometer. But Eve cannot *perfectly* imitate Alice, since she *does not know* whether or not Alice has used an extension. She can only insert an extension at random or choose never to add one – whatever she does, she will not have imitated Alice correctly in half the cases.

Let's discuss in particular the case in which Alice and Bob have not added an extension, whereas Eve has added one: the content of the *Extension* column is identical for both Alice and Bob, therefore their bit should be identical. However, Bob has not received the particle sent by Alice, but that sent by Eve, and Eve has not imitated Alice correctly. So at Bob's end, the particle can take the wrong output. By noting all of the possibilities, the reader can easily verify that Bob's bit and Alice's bit are different in 25% of cases (whereas they should always be the same). Such a percentage of error is seen immediately in the verification step – Alice and Bob can see the presence of an eavesdropper on the quantum line.

5.4 The fruits of an idea

5.4.1 From Bangalore to Geneva

Gilles Brassard ends his lecture. The impact has been lukewarm. Some audience members have found the idea pleasant, nothing more. Others are proving more reserved, in view of the technological abyss that, they think, separates the idea from its realization. Most have filed this lecture away among the trivial concepts. We must wait several years before research into quantum cryptography literally explodes[36]. In 1991, without being aware of the previous work, Artur Ekert, then a Ph.D. student at Oxford University, proposes his version of quantum cryptography. Bennett and Brassard, along with N. David Mermin, note that Ekert's protocol is equivalent to the one that they had presented at Bangalore. That could have provoked a schoolyard fight – very happily, instead of entering into a conflict, Bennett, Brassard and Ekert begin collaborating. In their enthusiasm, they train Asher Peres, Nicolas Gisin, Chris Fuchs and many others.

Elsewhere, experimental achievements begin to take place on the laboratory tables at IBM and British Telecom. In 1996, the quantum optics group at Geneva achieves the first quantum key distribution outside the laboratory – Alice is in Geneva, Bob is in Nyon, some 20 km away. The photons are sent via standard optic fibres, which normally maintain telephone links. Since then, the lectures of the members of the Geneva group are never without a photo of the 'laboratory' – a magnificent aerial view of Lake Geneva!

5.4.2 A change of perspective

It is difficult to predict the future of quantum cryptography. On the one hand, even if this method of key distribution is *physically* secure, the methods currently used by the armed forces, secret services and even in civilian life (payments over the internet, bank transactions), are already very secure in practice. On the other hand, we cannot limit modern cryptography to secret key protocols only, and for other fundamental communication protocols it is proven that quantum physics does not bring any advantages. Finally, quantum cryptography is as vulnerable as other methods to the oldest espionage technique – corruption of one of the authorized parties!

But it is well known that eminent scientists had voiced reservations concerning the usefulness of the transistor or the computer – there will always be pessimists, but it is the enthusiasts who (sometimes) change the way things are. My colleague Grégoire Ribordy, who, after finishing his thesis, established a little enterprise with the aim of producing in particular quantum cryptography apparatus, is certainly part of this last category.

Be that as it may, Bennett, Brassard, Ekert and the others have already made a huge contribution to physics. In effect, since the beginnings of quantum physics, the majority of physicists have presented the surprising aspects of the behaviour of particles as *limitations* – we cannot know by which path a particle travels, we cannot make a measurement without disrupting the results of successive measurements . . . Quantum cryptography does not repudiate that, but *it benefits from it* – since any measurement disrupts the results, the eavesdropper is going to be detected as soon as they try to make a measurement, to get information. It is a radical change of perspective – quantum physics is not an imperfect physics, but a *new physics*, which allows the achievement of tasks that are otherwise impossible[37].

Part 2
Quantum Correlations

·6·
Indistinguishability at a distance

6.1 Saint-Michel, second lecture

During my first lecture at Saint-Michel College, I had sidestepped the question about randomness that the student had asked. As the reader will recall, it was the experiment of **Fig. 1.1**, as simple as it was, in which the particles arrived one after the other at a beam splitter. I had said that we do not know how to describe the behaviour of each particle, we do not know why a given particle is reflected rather than transmitted. Is it really uncertain, random (which is the current opinion among physicists) or is there a mechanism yet to be discovered? I had deliberately turned attention from this problem, in order to concentrate on the experiment of **Fig. 1.3**, the Mach–Zehnder interferometer. We have seen, then, that quantum randomness is a curious randomness – if we assemble two 'randomness generators' (beam splitters) in a certain way, we rediscover the certainty! I briefly remind the assembled students of this, a week later, in the old granary of the College.

After having left the College the week before, we have undertaken a vast survey that has allowed us to familiarize ourselves with quantum interference, and to discard certain interpretations that, at first glance, seemed like good solutions. There seems to be no mechanism behind the astonishing quantum behaviour, and it is not clear whether or not there is a qualitative transition between the quantum and the classical world. To crown it all, we have seen that this bizarre randomness, difficult as it is to interpret, still turns out to be useful for carrying out certain operations, such as in secret key distribution. I think we will agree that quantum physics is much more than mechanics tinted by randomness. We are in the presence of a mode of functioning in nature that was unexpected. This mode of functioning was summarized in the indistinguishability principle, which we were able to reformulate like this: the behaviour of quantum objects depends on all of the indistinguishable possibilities.

In the course on the interference of a particle with itself – the students, like the reader, recall – we made an 'intuitive' prediction, which turned out to be incorrect in the experimental test. This time I am choosing the opposite route – we are going to *start with the indistinguishability principle, apply it to a physical system made up of two particles, and see what the predictions are,* then we will examine if and how these predictions are surprising. For the reader, this analysis will be pursued in Chapter 7, and the experiments that were actually performed in the laboratory will be discussed in Chapter 8. The phenomenon presented here (two-particle interferences) does not exhaust the whole of quantum mechanics, but will provide enough insight to discuss the interpretations without getting stuck in the wave-particle duality; this will be the content of Chapter 9.

6.2 Two-particle (in)distinguishability

6.2.1 The Franson interferometer

The interferometric apparatus that we are going to study is illustrated in **Fig. 6.1**. This time, the source emits *pairs* of particles.

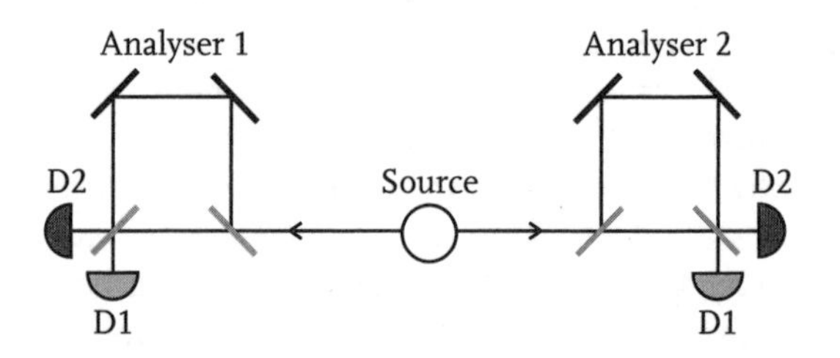

Fig. 6.1 Franson interferometer for two particles, balanced.

One particle goes to the right, the other to the left. Each particle encounters an interferometer, which at first glance resembles the Mach–Zehnder apparatus, but that comprises one important difference – the two paths are of very different lengths. This two-particle apparatus is called the *Franson interferometer*[38]. It is clear that four alternatives exist:

1. Long-Long (LL): both particles take the long path;
2. Long-Short (LS): the particle on the left takes to long path, that on the right takes the short path;
3. Short-Long (SL): the inverse of the preceding alternative;
4. Short-Short (SS): both particles take the short path.

In order to apply the indistinguishability principle, we must establish whether at least two of these alternatives are indistinguishable. If all of these alternatives were distinguishable from one another, we would not expect any interference.

It is evident that LS and SL are distinguishable from one another, and from both LL and SS. Indeed, let's take the alternative LS: in this case, the particle on the left reaches its detector well after the particle on the right reaches its, because the particle on the right has taken the short path while the other has taken the long path. In the alternative SL, it is the opposite – the particle on the left reaches its detector well before the particle on the right. Therefore by measuring the detection time we can distinguish LS and SL. But what about LL and SS? In these two alternatives, the particles arrive at the detector at the same time.

Casually, one of the students interrupts me without even raising a hand, 'Yes, but if the two particles have taken the long path, they arrive later than those which took the short path, so the two paths are also distinguishable from one another'. That is fair, I admit... provided that we know the moment in which the particles are emitted. The Franson interferometer is only relevant if the source *does not allow* the moment of emission to be known, and this puts some requirements on the source that I am not going to discuss in this text[39]. This issue being resolved, we arrive at the conclusion that the alternatives SS and LL are indistinguishable – we can expect to observe an interference phenomenon.

6.2.2 The interference phenomenon

In order to approach the interference characteristics of two particles, it is a good idea to begin with the following remark: the two alternatives SS and LL are intrinsically the alternatives *of two particles*. Indeed, the condition of indistinguishability is satisfied in the apparatus when the two particles have taken a path of the same length. Now, we cannot know that two particles have taken a path of the *same* length if we only look at one particle – the word 'same' implies a comparison. Therefore, the indistinguishability is apparent for the pair of particles, and not for each particle alone. But if such is the case, *the interference is also only apparent when we observe both particles.* And indeed, for the apparatus in **Fig. 6.1**, quantum physics predicts that, whenever the particles are detected at the same time: if the particle on the left is detected at D_1, the particle on the right is also detected at D_1; if the particle on the left is detected at D_2, the particle on the right is also detected at D_2. It should be stressed that both the left and right particles arrive sometimes at D_1, sometimes at D_2, with a probability of 50% for each alternative – we observe no interference for one particle[40]. But when we compare the results, we notice that *both particles have given the same result each time.* We speak of a perfect *correlation.*

Then, as in the case of one particle interference, it is a question of varying the length of one path or, more generally, of introducing a modification in one of the indistinguishable alternatives. Here, it is enough, for example, to lengthen the long path for the particle on the right, as in **Fig. 6.2** (naturally, when both particles have taken the respective long path, the difference in length introduced must be small enough, otherwise the particle on the right will arrive clearly after that on the left and LL would become distinguishable from SS). If we make this modification, it produces an effect similar to that which we had seen in one particle interference: a certain difference in the paths exists for which the predictions are inverted, namely, if the particle on the left is detected at D_1, then the particle on the right is detected at D_2; if the particle on the left is detected at D_2, then the particle on the right is detected at D_1. As previously,

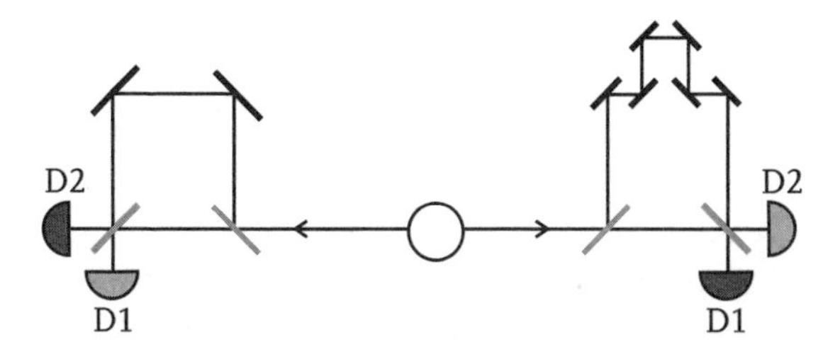

Fig. 6.2 Unbalanced Franson interferometer.

both the left and right particles arrive just as often as each other sometimes at D_1, sometimes at D_2, with a probability of 50%, but this time, when we compare the results, we notice that *both particles have given an opposite result each time.* We speak of perfect *anti-correlation.*

In summary, two-particle interference consists of this: by varying one of the alternatives, we can change perfect correlation to perfect anti-correlation for the measurement of the pair of particles. For other path lengths, we find intermediate correlations between the two extreme cases.

Before examining the implications of this prediction, I would like to mention to the reader that the Franson interferometer does not have an analogy for classical waves, whereas, as we have seen in Chapter 2, the Mach–Zehnder and Young interferometers were conceived initially in the framework of classical optics. The reason is that *classical waves do no obey a principle of indistinguishability.* This comment shows how the notion of 'wave-particle duality', forged at the beginnings of quantum physics, has dated, to the benefit of more general notions such as indistinguishability or Bohr's complementarity. In fact, only the advocates of the interpretation known as 'pilot waves', which we will briefly encounter in Chapter 9, can continue to describe quantum behaviour as a derivative of a duality between wave and corpuscle – a wave, of course, with exceptional characteristics, since it must bear everything surprising in quantum physics. Let's return now to our analysis of two-particle interference.

6.3 First exploration of the consequences

6.3.1 The principle and the surprise

Should we be surprised by the interference? More precisely, should we be more surprised by two-particle interference than by single-particle interference? In a certain sense, the answer is *no* – both types of interference have already been summarized by a single principle, the indistinguishability principle. So, one could say – and in fact some physicists (fewer and fewer) do say – that two-particle interference does not bring anything fundamentally new to quantum physics.

Nevertheless, I disagree with that position, as many other physicists (more and more) do. My personal opinion is quite strong, namely: I have come to think that the ultimate foundations of the indistinguishability principle are two- and more-particle interference. It is not yet time to discuss the reasons for my preferences – let's rather start by looking at the phenomenon of two-particle interference and its consequences[41].

For the single-particle interferometer, we have noted that each particle is aware of the modifications that we have introduced on only one path, and we were forced to admit that quantum particles 'explore' all of the paths, or 'are informed' about all of the paths. Let's look now at **Fig. 6.2**: the modification is on only one path and *for only one particle*. Therefore the resulting delocalization through the indistinguishability principle is even more dramatic. First, the interference effect is produced when both particles have taken a path of the same length (LL or SS), but how does the particle on the right 'know' that its companion on the left has taken a path the same length as it has taken? Secondly, how do the particles 'know' that the respective paths had *almost* the same length, this little 'almost' being enough to transform the perfect correlations into perfect anti-correlations?

We can propose a response to these questions that seems completely natural: one particle, for example, the first to arrive at its detector, *sends a message* to its companion informing it of everything it has encountered; and in taking this information into account, the other particle is going to arrive at the appropriate detector. However, if we believe in the indistinguishability principle, the interference must be manifest *no matter what the distance* between the two particles – the argument by which we have proven the indistinguishability between the alternatives LL and SS remains valid, whether the distance between the detectors is a millimetre or a light year! Moreover, it appears well established in physics that a speed limit exists for the sending of information, that is, the speed of light. Therefore, if the detectors are very far apart, it is impossible that a message sent at a slower speed than that of light could be sent from one particle to another and arrive in time to influence the behaviour of this second particle.

6.3.2 Three explanations (at least)

We now find ourselves at a crossroads, because the three following points are incompatible: (i) the indistinguishability principle is correct and complete; (ii) there is an exchange of information between the particles; (iii) a speed limit for the propagation of information exists. Which of these points must be abandoned?

1. If we consider that the correlations between two particles are necessarily due to the sending of a message, and that we want to retain the speed of light as the speed limit, then the indistinguishability principle is incomplete. It would be necessary to complete it by saying that the interference is certainly manifest as soon as two alternatives are indistinguishable, but then only provided that the distance between the particles permits the sending of a message.
2. As for the quantum correlations, we can abandon the characteristic of the limit of the speed of light, because – as I will explain in detail in the next section – two physicists cannot use these correlations to 'communicate'. Therefore we can postulate that the particles can send messages faster than the

speed of light, as we are not in any way able to profit from doing the same.

3. Finally, we can abandon point (ii) – give up the explanation of quantum correlations by an exchange of information. In this case point (iii) also loses its relevance – we remain with what we had: the indistinguishability principle.

Before giving the floor over to the experiment, I would like to analyse the above three possibilities in a little more detail – after all, I was invited to this College by a professor of philosophy.

Possibility 3, that is the indistinguishability principle without any underlying mechanism, is just quantum theory. Possibility 1 is interesting, because it can be subjected to the test of an experiment – it 'suffices' to put a sufficiently great distance between the detectors. If the two-particle interference disappears, the indistinguishability principle is generally proven false, and must be completed by a rule about distances. We will discuss the experiments in Chapter 8, but let's affirm now that the two-particle interference does not disappear, even if the distance becomes greater than that at which light is able to travel in the interval between the two detections.

As for possibility 2, it is quite complex to discuss its validity. In effect, we postulate a *'hidden'* (impossible to observe) *supraluminal communication* between the particles. At the moment when we put such a hypothesis to the test of an experiment, the physicist is forced to add a number of supplementary working hypotheses – remember Chapters 3 and 4: it is not easy to conceive an experiment; and in order to set out these hypotheses, there is no other guide than the imagination (we are talking about something unobservable!). Since in any case the existence of this hidden communication is perceived as extremely improbable by the majority of physicists, the efforts made to explore this possibility are minimal. I am not an advocate of hidden communication and I consider that there needs to be nothing more done with it for the moment, but in all honesty, neither can I make the reader believe that this hypothesis has been proven false according to the criteria of experimental science[42].

6.3.3 Sending a message?

I have already subjected the students to a considerable *tour de force* – we have reviewed the indistinguishability principle, introduced the notion of correlation and revealed interference in correlations. Then, we saw that this prediction of quantum physics seems to throw back into question a well-known and well-established fact in physics, the fact that the speed of light is the limiting speed for communication. I cannot let it rest there, even if my audience is tired. I cannot depart leaving those who are listening to me with the impression that quantum physics could one day allow communication faster than light[43].

Whatever the explanation might be, the quantum particles introduced correlations at a distance. However, *this phenomenon cannot be employed for communication,* it cannot be used to send a message, whether faster or slower than light. The reason for this is: whether we are in a situation of perfect correlation, perfect anti-correlation, or whatever situation in between, concerning the correlation of two particles, *nothing changes in the results that we observe for each particle individually.* Specifically, for the Franson interferometer that we have considered, we have said that for each side, half the particles are detected at one detector, the other half at the other. Alice, who observes only the particles that have gone to the left, sees random detections; on the right, Bob may modify his interferometer at will and nothing will change for Alice. It is only when Alice and Bob speak to each other (by telephone, for example) and they compare their results, that they notice the existence of correlations between the particles. An ordinary medium of communication (telephone, internet, meeting at a bistro) is therefore absolutely necessary in order to be aware of the quantum correlations – these correlations with only themselves do not allow communication.

6.4 Continuation of the programme

The issue of communication being clarified, I tell the students quickly that the correlations at a distance between two particles have been experimentally verified, but that they still constitute an area of research, and I go on to a brief outline of the interpretations of quantum physics.

With the reader, we will approach the interpretations in Chapter 9 and we are going to continue to examine two-particle interference in the next two chapters. I propose to the reader the same itinerary that we followed in Part 1, namely: an initial chapter devoted to a more abstract presentation of the notion of correlations and several historical elements (Chapter 7); then a further chapter devoted to the experiments that have effectively demonstrated these correlations and to the surprising criticisms that have been voiced against these observations (Chapter 8).

·7·

On the origin of correlations

7.1 The Bell theorem

In Chapter 2, we saw that indistinguishability deriving from cars in relation to the knowledge of a pedestrian does not lead to any surprising modification in the properties of those cars. Indistinguishability in everyday life is linked to the ignorance of the observer, which can be overcome, while indistinguishability at the particle level cannot be removed without leading to qualitative change of the physical properties. The arguments in Chapter 2, as well as the failure of the 'Heisenberg mechanism' discussed in Chapter 4, have taught us that quantum indistinguishability is not a kind of insurmountable ignorance. Quantum indistinguishability is therefore a new idea – to affirm that two paths for a car are indistinguishable is not the same as affirming that two alternatives for a particle are indistinguishable. We are going to see that it is possible to argue in a similar way on the subject of correlations – in the quantum world, the notion of correlations appears in a new, unexpected light.

7.1.1 Referees, pastrycooks and particles

In everyday life, we know of two ways of establishing correlations:

1. Correlation by *exchange of signal*. In a football match, every time the referee gives a signal (a blow on a whistle) all of the players stop.
2. Correlation *established at the source*. I order from a pastrycook two identical gift boxes to send to two people who know each

other, to avoid envy. If one of the people finds a chocolate cake in their box, we know with certainty that the other will find a chocolate cake in theirs.

Now, we have seen in the preceding chapter that correlation by exchange of signal is problematic in the case of quantum particles – quantum physics predicts that the correlations will be observed regardless of the distance that separates the particles at the moment when they arrive at the detectors, while no signal can be propagated faster than light[44]. Could it be then that the correlations of the pairs of particles are *predictable at the source*? This would appear to be the only 'reasonable' alternative: correlations should derive from some information (called *local variables*[45]) that the particles share before flying away from one another. However, in view of scientific methods and what we already know about quantum physics, we would like to not just be content with our intuition but to *test* the hypothesis in question. But how to do it? In order to discard the possibility of an exchange of signal, as we have just said, it is necessary to separate the particles by a sufficiently large distance. But what type of experiment will it be necessary to conceive to test the hypothesis that the correlations are established at the source? I would like the reader to think about this question for a few seconds, before reading on.

In fact, conceiving the said experiment is not obvious. The response that I am going to offer is very clever and can only be discovered by somebody who knows how to work with quantum theory. This type of argument is known as the *Bell theorem*, after the physicist who first proposed it, John Bell. The Bell theorem can be described with the tools of elementary mathematics – the following paragraphs are devoted to this approach. This is the most difficult part of the book. The reader can skip it and keep in mind simply that *the Bell theorem furnishes a criterion to experimentally exclude the hypothesis that quantum correlations are established at the source (that is, to rule out an alternative description of quantum phenomena based on local variables)*. I find, however, that the effort of understanding the Bell theorem is worth it, because there are few

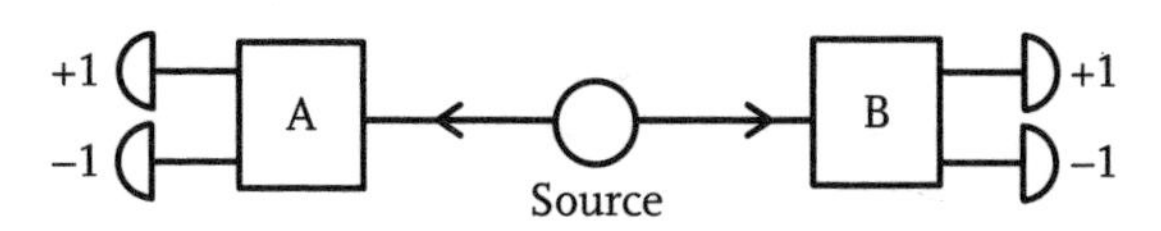

Fig. 7.1 Diagram of the principle of a two-particle measurement: The property A is measured for the particle of the left, the property B for that on the right.

other profound results in modern physics that can be explained with elementary concepts.

7.1.2 The Bell theorem: preliminary notions

We are considering the experimental apparatus of **Fig. 7.1**, which generalizes the Franson interferometer. A source emits a pair of particles likely to show quantum correlations. One particle goes towards Alice's laboratory, the other towards Bob's. Each of these two physicists subjects the particle they receive to the measurement of their choice (in the example of the Franson interferometer, that could mean that Alice and Bob can choose to add a small extension to one of the paths on their side, as in **Fig. 6.2**). We want to test the following hypothesis:

Hypothesis: *the correlations between the two particles are established at the source.* In other words, before leaving the source, the particles share all of the common information needed to produce the correlations (the local variables).

In order to go on with the argument, we must note that our hypothesis naturally leads to the following consequence:

Consequence: *the correlation established at the source between the particles must not depend on the measurement the physicists have chosen to effect on each of the particles.*

We can express this hypothesis in a more anthropomorphic way by saying that at the moment of leaving the source, the particles do not 'know' to what kind of measurement they will be subjected. This consequence is natural, because the choice of a measurement (of whether or not to add the extension, in our example) may be made by Alice and by Bob *after* the particles have left the source. Let's note in passing that, for this argument to make sense, we must assume that the physicists can make free choices – or at least, choices that are uncorrelated to any meaningful parameter of the particles. If every action, even human, is pre-determined, then the Bell theorem is not valid – but in fact, if everything is pre-determined, there is no further need to look for an explanation for quantum correlations or for anything else!

Before going on to the demonstration of the Bell theorem, we can add a few supplementary conditions:

Condition 1: for each pair of particles, Alice and Bob each choose one possible measurement *out of two*. In concrete terms, we will call Alice's measurements A and A′, and Bob's B and B′. There are therefore four possibilities for the measurements on the pair: (A,B), (A′,B), (A,B′) and (A′,B′).

Condition 2: each of the measurements A, A′, B and B′ is a measurement with two outcomes, or *dichotomic*. We have already worked a great deal with dichotomic measurements, for example, the apparatuses of **Figs 1.1**, **1.3** and **1.4**, in which the particle is detected on one of two possible paths. In the Franson interferometer presented in the preceding chapter, each particle is equally subjected to a dichotomic measurement. More abstractly, we can think of the measurements A, A′, B and B′ as boxes fitted with two indicator lights, let's say one green light and one red light: each time a particle arrives, one of the two lights (but never both at once) is illuminated.

The reader should note that these two conditions *simplify* the problem – in fact, they reduce it to its minimal complexity[46]. It is notable that the Bell theorem can already be demonstrated in this case.

To finish, we must establish a convention. For each measurement A, A′, B and B′, if the green light is illuminated we will say that the result of the measurement is +1, if the red light is illuminated we will say that the result is −1. This choice of numbers is completely arbitrary, we could have chosen +37 and +3.1415 for the two results – the Bell theorem is simply expressed and demonstrated more elegantly with our choice.

7.1.3 The Bell theorem: statement

We can now attack the Bell theorem. Let's consider a given pair of particles. According to our hypothesis, each of these particles leaves the source with certain information – to cement the ideas, the reader can think of both particles as leaving the source with the same information, although this constraint is not necessary for the argument that follows. Now, we accept that the particles do not communicate between themselves after having left the source. Consequently, everything that happens on the route between the source and the detectors can only reduce the correlations, whose origin (according to our hypothesis) is only the information shared at the source.

Given this, we can suppose that the information given at the source determines, in fact, *the result of each possible measurement*[47]. The list of results (or, equally, the information from which they are computed) is the local variable. It is therefore determined at the source that for the pair that concerns us, if Alice measures A she is going to obtain the result a, if she measures A′ she will obtain the result a', and the same on Bob's side. Remember that a, a', b and b' are either valued at $+1$, or at -1.

Specifically then, each pair of particles carries with it information sufficient to calculate the following number: $S = (a + a')b + (a - a')b'$. For the reader, this number is a godsend. In any case, it is not a very difficult number – in fact, for each pair of particles it can only have a value of $+2$ or -2. To see this, it is enough to note that: (i) if $a = a'$, the second term of the sum is zero; as for the first term, $(a + a')$ is either valued at $+2$, or at -2, and b is either valued at $+1$, or at -1; (ii) if $a = -a'$, it is the first term that is zero, and for the other we apply the same reasoning.

We note that, for a given pair of particles, Alice and Bob cannot measure the value of S, since Alice sometimes measures A, sometimes A′, and if she measures A she has no idea of the result that could be obtained in measuring A′; and the same for Bob. However, in making different measurements for a large number of pairs, *they can measure the average value of S*. To see this, it suffices to rewrite S in the form $ab + a'b + ab' - a'b'$: the average value of S then is the algebraic sum of the four average values corresponding to the measurements that we can actually make, namely (A,B), (A′,B), (A,B′) and (A′,B′). So, experimentally, we have access to the average value of S. Now, here is the statement of the Bell theorem: *if our hypothesis is correct, the average value of S must be a number from −2 to +2 inclusive.*

This is not a great discovery – in fact, it is absolutely trivial! Just as a student from the Saint-Michel College, where the scale of grades is 1 to 6, cannot end the year with an average of 7, likewise the average of the number S, for which each outcome is either valued at +2, or at −2, cannot be outside those boundaries. The great discovery is this: for two suitably prepared particles and for certain well chosen A, A′, B, B′ measurements, quantum theory predicts that the average value of S is double the square root of 2, namely closer to 2.8284, and that is *greater than* 2, indisputably! The Bell inequality, that is those easily derived boundaries for the average value of a certain number, is *violated* in the framework of quantum theory. Since the demonstration shows no grey areas, we are forced to conclude that the hypothesis at the point of departure is not compatible with quantum theory.

This result is exactly the criterion for which we were searching. We have put forward the hypothesis that quantum correlations could be established at the source, given that it is difficult to accept that they are due to the exchange of a signal. This hypothesis implies that a certain *measurable* quantity, the average value of S, must be less than or equal to 2. Quantum theory, on the other hand, predicts that this quantity can have the value $2\sqrt{2}$. It no longer remains for us to measure this number in the laboratory – if the value measured is greater than 2, we shall know that quantum correlations cannot be established at the source.

7.1.4 Commentaries on the Bell theorem

This ingenious and significant result has been the subject of countless discussions. I want to present to the reader several additional commentaries on the subject, without claiming to be exhaustive.

First, we already knew from Chapters 1 and 2 that the results of the measurements of one quantum particle depend on the experimental apparatus, for example, we have seen that it is not possible to measure by which path a particle has travelled in a Mach–Zehnder interferometer, without significantly modifying the outcome of the measurement (that is, on which path the particle is detected). Now, the hypothesis that quantum correlations are established at the source drove us to accept that, for each pair of particles emitted, the result of *any* measurement on each particle is determined at the source. Isn't there a contradiction in principle between this hypothesis and the observations on one particle? It turns out that there is no contradiction: *one can reproduce experiments on single quantum particles with suitable local variables*! For all of the interferometers of Part 1 of this book, Bohm's theory (see Chapter 9) does the job pretty well; for a spin, it was John Bell himself who found a local-variable model – incidentally, this was believed to be impossible because of a theorem by von Neumann; the theorem was mathematically correct, but Bell noticed that one of the hypotheses of the theorem was too strong and unnecessary, and by removing it the restraint no longer held. The existence of such local-variable models is the reason why I have stressed many times in Part 1 that single-particle phenomena and their wave–particle duality are not enough to fully appreciate quantum physics; one must study two- and more-particle phenomena as well.

Second, in order to do justice to a large number of physicists, we must point out that *many Bell-type theorems* exist. In concrete terms, the construction of S that we have adopted is not that originally proposed by Bell[48], but it is due to four other physicists, namely John Clauser, Michael Horne, Abner Shimony and Dick Holt, and it is therefore known as CHSH[49]. I chose this construction because it is the one that, in my opinion, allows the theorem to be proven in the most immediate manner. But it is a question of taste – in his celebrated articles for the general public, Mermin preferred to present a Bell theorem that involves *three* dichotomic measurements for each particle[50]. Bell theorems exist that involve more measurements for each particle, but also that use non-dichotomic measurements, or even that concern correlations between a greater number of particles. All of these theorems coincide in their spirit – we suppose that the correlations are established at the source, we formulate an inequality ('S must be smaller than or equal to 2'), then we show that *quantum theory predicts that the inequality can be violated.* Moreover, for three particles or more, there also exists another type of theorem[51], which is not based on inequalities, and that allows a contradiction to be demonstrated between quantum theory and the hypothesis that correlations are established at the source. The idea for this other theorem come from Daniel Greenberger, who worked on it in collaboration with Horne and Zeilinger, whence comes the name GHZ. It is equally easy to demonstrate and may constitute a good exercise for the reader who would like to delve further in this direction.

Third, I must mention here two terminological questions. The term *non-locality* appears very often in the discussions on the Bell theorem; some people don't like it, probably because it suggests that some communication is going on. Curiously, a much more problematic expression is well accepted: to express the fact that the preparation at the source is not enough to explain quantum correlations, people used to say that quantum physics is incompatible with *local realism.* This expression is problematic because it is equivocal to everyone who is in contact with the world of philosophy[52]. Anyway, as in every terminological question, both terms are ultimately innocuous as long as we know to what they refer – obviously, they can give rise to some absurd deviations if they are taken out of context.

In summary, Bell's idea has given rise to lively activity in theoretical physics and a large number of debates on the interpretation of quantum physics. But the main point is, the Bell criterion is quantitative – it will allow the experimental resolution between the hypothesis that correlations are established at the source and the prediction of quantum theory. Before we turn our attention to the experiments that have been performed, we shall quickly review the history of the debate over quantum correlations.

7.2 Brief history of quantum correlations

7.2.1 Einstein–Podolski–Rosen and non-locality

In Chapter 2, we talked about the first clues that put physicists onto the path of quantum mechanics – that was in the very first years of the twentieth century. It is only in 1926 that the basis of the new theory, just as we practice it today, is set out through the works of Heisenberg and Schrödinger. Science historians speak of an 'old quantum theory' when referring to all of the partial results obtained before 1926, such as the blackbody-radiation spectrum calculated by Planck, or Bohr's model for the hydrogen atom. Also belonging to this old quantum theory are two great contributions from Einstein – the explanation of the photoelectric effect in terms of light quanta (later called *photons*) and the study of the emission of light by de-excitation of an energy level, the basic principle of laser physics.

But as soon as the new quantum theory is set out in a concrete form, Einstein openly declares his scepticism. He does not deny the temporary value of the new theory, but he refuses to accept that what we have before us is the final version that really best describes natural phenomena. Over many years, Einstein is going to attempt to find the flaw. It is a dynamic task, a matter of attempts, abandonment, renewed attacks[53].

Situated within this context is the article[54] written in collaboration with Boris Podolski and Nathan Rosen in 1935 and that will be known as the *EPR argument* – sometimes, but incorrectly, called the EPR paradox. This article is the first work in which it is noticed that quantum theory predicts correlations that are instantly established between two distant particles. For the creator of the theory of relativity, constructed on the principle that all communication is propagated at a speed slower than that of light, such a prediction can only be an artefact of an incomplete theory. Einstein thus identifies the flaw in the new quantum theory in the fact that it predicts a form of *non-locality*, a correlated behaviour of two objects that cannot communicate between themselves.

7.2.2 Schrödinger and non-separability

The non-locality of the EPR argument is convincing, because it touches on the notions of space and time. This argument provokes an immediate response from Bohr[55], who defends quantum physics in the name of his principle of complementarity. Also, Erwin Schrödinger, sceptical like Einstein with regard to the new quantum physics and in search of the flaw, is turning his attention towards the quantum description of systems composed of several particles. Schrödinger puts his finger[56] on a problem that is more difficult to describe than non-locality, but that has profound ramifications for our vision of the physical world. He noted that quantum correlations are associated with a notion of *non-separability*. The technical description is: two quantum particles can find each other in a state such that only the properties of the pair are defined. Let's try to gain a deeper understanding of this statement.

We have two particles – they are indeed two 'masses', which we can separate by a large distance as we have seen. This is important: Schrödinger's non-separability is neither a sort of fusion of the masses, nor a sort of chemical liaison (that is to say, the establishment of an energetic link that makes separation of the masses very difficult). The non-separability between the two particles is manifest at the level of the properties and goes as follows:

- *The set of two particles possesses well-defined properties,* for example, the spin of particle A and that of particle B go in opposite directions. The property 'being opposed' is clearly a property of two objects – we have already seen a similar property: 'arriving at the detector at the same time' in the Franson interferometer.

- On the other hand, *each particle taken separately possesses no well-defined property*: specifically, if I measure the direction of the spin of particle A, I can find it pointing in any direction.

There is no such situation in everyday life: when we say 'two cars are moving in opposite directions', we obviously mean that each car is moving in its own, well-defined direction, and that these two directions happen to be opposite to one another. By specifying only the fact that the directions are opposite, in the classical world we neglect part of the information (we could have specified which direction was which, and get the property of their being opposite as a mere consequence). In quantum theory, however, one can define a 'pure relative property', which is not a consequence of individual properties. This is related to the impossibility, pointed out in Chapter 2, of a description of properties in terms of set theory[57]. In order to explain more clearly what he was getting at, Schrödinger invented his famous argument of the cat[58].

Schrödinger has attached an adjective to non-separable states: he calls them *verschränkte*, a word that is used in modern German to indicate folded arms. In English, this is translated as *entangled*. The reader who browses scientific journals will have encountered this adjective – *entanglement* is one of the most important research subjects at the time of writing this book.

7.2.3 Thirty years on the shelf

When two quantum systems are correlated, it becomes impossible to describe them separately (Schrödinger), whatever the distance that separates them (EPR). But are these predictions *correct*? Does interference between two particles separated in space really exist? And do entangled states exist? These look like (and in my opinion are) essential questions . . . that, however, have been sidestepped, relegated for more than thirty years to the ranks of problems of interpretation, which is a synonym for 'useless speculation' for many physicists.

Historically, this fate is hardly surprising. Quantum physics in 1926 had opened up enormous prospects – it allowed the prediction of the characteristics of molecules (that is, the possibility of tackling the whole of chemistry in physical terms), the existence of a considerable number of 'particles' as well as the manner of observing them, the electrical and thermal properties of solids . . . It also allowed for the prediction of the possibility of the atomic bomb, and it is a mystery to nobody that a good number of great physicists contributed to the development of this weapon, if withdrawing later in view of its devastating effect.

After the war, the physics world is polarized like the political world. The Western scientific community revolves henceforth around the United States, *Physical Review* replaces *Annalen der Physik* as the most important scientific journal. An atmosphere takes hold that is referred to as the time of '*shut up and calculate*'. There is no time for 'sit down and contemplate'. Abner Shimony (whom we have already met as the 'S' of the CHSH argument) is working on his doctoral thesis in the 1950s. One day, to occupy him, his supervisor gives him the EPR article to read, with the instruction, 'Read this for me and find me the fault'. The insinuation is significant: the EPR argument must be false since it contradicts the theory – a theory, it is true, whose successes accumulate by the day. Shimony reads the article and is fascinated by it[59]. Since then, he has not ceased to be interested in correlations between separated particles.

Entanglement will free itself from the maze of interpretations to get closer to the laboratory mostly thanks to the work of John Bell. But before talking about Bell, this brief history of the EPR argument brings us to our first meeting with David Bohm. Bohm's name is known by most physicists through an ingenious one-particle interference phenomenon that he predicted with his student Yakir Aharonov, and that naturally bears the name Aharonov–Bohm. Most physicists, on the other hand, have not heard of the interpretation of quantum mechanics proposed by Bohm, because it does not conform to the orthodox doctrine, and does not therefore have the right to be mentioned in institutional courses. We will talk about it in Chapter 9, because Bohm's 'mechanics of pilot waves' is the most elaborate alternative interpretation, and it is highly instructive for clearing up its problematic aspects. Here, we are concerned with Bohm's contribution to the EPR argument. This contribution is rather technical in nature – Bohm rewrote the EPR argument in terms of two particle spins, whereas Einstein, Podolski and Rosen had used dynamic variables of position and momentum. It is an important step, because in the mathematical formalism of quantum physics the spin is the easiest system to deal with. This simplification opens the way for the work of Bell.

7.2.4 John Bell, the person

Serious and rather reserved, John Bell worked actively in a mainstream research domain (he was a particle physics theorist at CERN in Geneva), but he obtained his principal result by working on the 'philosophical' subject of quantum correlations. As we said above, the first step that he took consisted of removing the restraint of von Neumann's theorem, then by constructing an explicit local-variable model for single quantum particles. As a next step, he set out to find the local-variable model for two particles . . . and he came up with his own impossibility theorem. Is this theorem bound to fail as von Neumann's? This is highly improbable (in my view, utterly impossible): in his time, von Neumann's theorem was accepted almost without criticism; while Bell's theorem has already undergone forty years of intensive studies and has resisted any attack – moreover, as we saw in this very text, the formulation of the theorem is not difficult.

As for philosophical preferences, John Bell would have liked to find a local-variable model reproducing the whole of quantum physics: *a priori*, he favoured 'local realism'. But he honestly accepted the conclusion of his theorem and of the experiments. His premature death achieved his ascension to the status of cult physicist, apparent in the recollections of those who met him[60].

7.3 Return to the phenomena

We know now that for two particles the indistinguishability principle, that is, quantum theory, predicts correlations whose characteristics are the following:

1. quantum correlations do not disappear by increasing the distance between the particles, and therefore their origin cannot be the reception of a common signal;
2. quantum correlations violate Bell's inequality, and therefore nor can their origin be a common decision taken at the source.

In other words, if quantum theory is correct, neither of the two usual mechanisms that explain correlations can be invoked! But is quantum theory correct? Will correlations be maintained over a great distance? Are they really going to violate Bell's inequality? It is high time to return to the laboratory.

·8·
Orsay, Innsbruck, Geneva

8.1 The Aspect experiments (1981–82)

'Do you have a permanent position?'. The young Alain Aspect, visiting John Bell in Geneva, was speechless for a moment. Alain has an ambitious objective for his doctoral thesis – to perform, in the optics laboratory at the Paris-Sud University at Orsay, an experiment that allows him to measure correlations between two particles and to see if Bell's inequality is violated. An ambitious project. But the majority of the scientific community continues to regard Bell inequalities, non-locality, as 'philosophy'. John Bell, then, is expressing legitimate concern for the future of the young man who has come to talk with him, who risks finding himself marginalized if he goes on with his project. 'Do you have a permanent position?'. Yes, Alain has a permanent position, a small salary guaranteed for life – but he risks his prestige. The risk will pay off – Alain will perform his doctoral project and he will even have the pleasure of seeing the scientific community calling his experiment 'the Aspect experiment', with a worthy recognition that is not always guaranteed.

8.1.1 The first experiments

At the time when Aspect begins his thesis, several experiments in quantum correlations have already been performed, all in the United States. The first has been performed by Freedman and Clauser at Berkley (California) in 1972 – Clauser himself, then Holt at Harvard, and finally Fry and Thomson in Texas, have repeated the experiment several years later. The Bell inequalities appear to be violated in California and in Texas, but the experiment on the East coast seems to go against the quantum prediction. Interpretation of the results not being easy, doubt remains. The first two experiments[61] by Aspect, performed with Philippe Grangier and Gérard Roger in 1981, irrefutably confirm the observation of quantum correlations that violate Bell's inequality.

In all of these apparatuses, the indistinguishable alternatives do not concern the length of the paths taken as in the Franson interferometer[62], but a degree of freedom of the photons, *polarization*, which we can regard as similar to the spin that we have encountered for the neutron. The source produces two photons of *opposite polarization*, but none of the photons is polarized in any specific direction. It is a condition of two-particle indistinguishability ('being opposite'), as is usual if we want to observe interference in the correlations.

8.1.2 Locality loophole

The French physicists have therefore confirmed and improved on Clauser and Fry's observations – Holt must have made an error. However, the debate over non-locality is not closed yet. There are still two loopholes open, which prevent the conclusion that non-locality has been experimentally established. We call these the *locality loophole* and the *detection loophole*. We begin by discussing the first, and we will return to the second a little later.

The locality loophole consists of the following. We have seen that, according to quantum theory, pairs of particles establish correlations even at a large distance, when no communication between two particles *in the same pair* is possible at a speed less than that of light. But a correlation measurement is a statistical measurement – it is necessary to send *a great number of pairs of particles* into the apparatus, one after the other, in order to obtain a significant result. Now, if the elements of the apparatus do not change over the course of the experiment, then, after the time required for light propagation, information is, in principle, available about the way in which the physicist has arranged the interferometer. From then on, the following pairs may be informed (they might have received a message), and they only have to behave according to quantum rules.

The reader is right to find this argument a little artificial, but must not forget what is at stake – we observe correlations that in the first analysis cannot be explained either by judicious preparation (because of the violation of the Bell inequalities) or by the exchange of a message (because of the distance). It is understandable then, to ask ourselves if the first analysis was complete and if there was not a possibility of going back to one of the traditional explanations. The Bell inequalities clearly being violated, it is necessary to look for something on the side of the transmission of a signal.

We have seen that, in order to observe a violation of the Bell inequalities, it is necessary to study the given correlations in four different configurations, which we have called (A,B), (A,B′), (A′,B) and (A′,B′). In all of the experiments mentioned above, physicists first measured the correlations for the first configuration, then changed an interferometer and measured the correlations for the second configuration, and so forth. Proceeding in this way, the locality loophole remains open. In order to close this loophole, it would be necessary to set about doing it differently – it would be necessary to modify the interferometers during the course of the experiment, while the pairs of particles are being emitted one after the other. If this modification is sufficiently rapid and is done in a random way, each pair of particles will encounter an unpredictable interferometer – it would achieve nothing for information about previous settings to be available. In other words, in order to close the locality loophole it is necessary to implement the idea that we have mentioned in the preceding chapter – the two experimenters Alice and Bob must choose their measurement after each of the particles has left the source, 'freely' – independently of one another, and also independently of any other significant parameter of the experiment.

In a third experiment[63], performed in 1982, Aspect and his collaborators (Jean Dalibard replaces Philippe Grangier) insert into the apparatus some devices allowing a rapid change in the analysers.

No change in the correlations is observed – the first setback for the locality loophole. It is this experiment that we usually call the *Aspect experiment.* This is what Aspect presented at Caltech, which triggered Feynman's suggestion. From that moment on, the majority of physicists begin to consider non-locality as an established fact. But the manner of implementing the rapid changes in the Orsay experiment was not really ideal, and it will be necessary to wait sixteen years before the locality loophole can be permanently closed.

8.2 Two other experiments, in 1998

In the years that follow the Aspect experiment, the observation of quantum correlations is confirmed by a number of experiments, performed by several research groups[64]. I move on directly to the two experiments performed in 1998 in Innsbruck and Geneva, which constitute, in a sense, the crowning achievement of this research.

8.2.1 The Aspect experiment carried to perfection

So we have jumped sixteen years and about a thousand kilometres to find ourselves in Innsbruck in 1998. In the city of the Golden Roof, we meet up again with the group of Anton Zeilinger, the whole of which is about to move to Vienna – where, the reader will recall, they will notably demonstrate interference for the large C_{60} molecules.

The experiment[65] performed by Zeilinger and his collaborators Gregor Weihs, Thomas Jennewein, Christoph Simon and Harold Weinfurter is the definitive version of the Aspect experiment of 1982. The photons emitted by the source – a source, moreover, of a different type and more efficient than that used in Orsay – travel along optical fibres installed in the campus of the University of Innsbruck, to the analysers, which are found at a distance of 400 *metres* apart (in Aspect's experiment, the whole apparatus was confined to a laboratory, therefore the distance between the analysers was a few metres). At such distances and with judicious electronics, it is possible to implement rapid and random changes that assure that each particle cannot be informed about the configuration that its companion will encounter. The correlations persist in violating Bell's inequality – the locality loophole is permanently closed!

8.2.2 Correlations at 10 km

The Austrians' article appears in the December 7th 1998 edition of the journal *Physical Review Letters.* A month and a half prior, on October 26th, another quantum-correlation experiment had appeared in the same journal[66]. It was by Wolfgang Tittel, Jürgen Brendel, Hugo Zbinden and Nicolas Gisin. The Geneva group is doing it again: those who in 1996 had demonstrated the feasibility of quantum cryptography over long distances (20 km) demonstrate two years later that quantum correlations are equally stable and violate Bell's inequality over distances of kilometres. While Zeilinger's group ran their own optical fibres through the university campus at Innsbruck, the Geneva group adopted another strategy – asking the Swiss telecommunications operator to be allowed to use, for several hours, the fibres already installed between the telecom stations. On the appointed day, the physicists distribute themselves between the stations at Cornavin (in the heart of Geneva), Bernex and Bellevue (two outer suburban areas). At Cornavin they put the source of the pairs of photons, and at Bernex and Bellevue the two analysers – it is a Franson interferometer. For the non-locality, what is important is the distance between the two analysis stations, Bellevue and Bernex – 10.9 km as the crow flies. The correlations violate Bell's inequality just as in the Innsbruck experiment, without any possible ambiguity.

The Geneva physicists have not added rapid switching to their experiment. Unlike that of Innsbruck, their experiment is not designed to close the locality loophole, but to demonstrate the violation of Bell's inequality over large distances. This experiment is probably the one that has caused the most excitement. When, in the year 2000, the American Physical Society wanted to record, in ten posters, the stages marking twentieth century physics, quantum correlations gained a place in these posters thanks to the Geneva experiment[67].

8.3 A curious argument

We have before us some experiments, reproduced by several independent research groups, which confirm the theoretical prediction: all of the criteria appear to be assembled so that we are able to conclude that quantum interference of distant particles is *confirmed experimentally*. It is in fact the conclusion drawn by the majority of physicists . . . and what objection could we still raise?

One objection has nevertheless been put forward, based on the imperfection of the detectors. Current photon counters have a fairly limited efficiency – they detect at best (let's be optimistic to simplify things) half of the photons. In order to understand the argument, which we call the *detection loophole*, I will begin with an example inspired by everyday life.

Let's suppose that police radars only measure the speed of half the cars. This could be due simply to the slowness of the electronics within the radar, which, after having measured the speed of one car, has a certain amount of dead time before being able to measure another. In this case, the statistics for offences are significant, all the same. But there could be another reason for the fact that the radar does not see half the vehicles – the police could have badly installed their radar, such that only vehicles that are tall enough send back a signal, so sports cars, always lower than average, are not seen. In this case, all of the sports cars can exceed the speed limit without being seen – the statistics for offences will be distorted, because we only measure the speed of the slower vehicles[68].

The detection loophole is based on the same idea. Current detectors detect less than half the photons that are sent. This is a fact, but as in the example of the cars, it is legitimate to ask ourselves whether or not the photons detected constitute a representative sample of all of the photons. It could be that it is not the case, that only *certain photons*, suitably 'programmed', activate our detectors. These photons, the detection loophole argument continues, could be, furthermore, programmed to violate Bell's inequality, but if we detected all of the photons, we might see that Bell's inequality is not violated.

In order to grasp the weirdness of this loophole, it is necessary to remember the sessions in the school laboratory or even at university. At one time or another, every one of us obtains an experimental result that *disagrees* with the theoretical prediction. We have looked for the error, and if we failed to find it, we have written in our report a loose statement like, 'the instruments are too imprecise'. We have cited the uncertainty of the measurements to explain the *disagreement* between the experiment and theoretical calculation. The detection loophole is perhaps the first example in the history of physics where the imprecision of the measurements is cited to explain the *perfect agreement* between theory and experiment!

Just as for John Bell, it is difficult for me to believe that quantum theory gives precise predictions only because of the poor efficiency of the detectors, and that it is destined for a miserable failure the day our detectors are perfect[69]. It is equally necessary to know that techniques exist (*ion traps*) in which the detectors are practically perfect, and the quantum correlations do not disappear. These experiments close the detection loophole, but unfortunately the particles (ions, that is, atoms that have lost or gained one or more electrons) are very close to each other, and therefore the locality loophole stays open[70]. At the time of writing this book, what is lacking in order to convince the last sceptics is an experiment in which *both* loopholes are closed. A few proposals exist, and it is possible that, by the time the reader reads these lines, this experiment will have been performed. The two loopholes of locality and detection will have disappeared from the scene then, last witnesses to the great discussions about two-particle correlations begun by the sceptic Albert Einstein and the orthodox Niels Bohr in 1935.

8.4 'Experimental metaphysics'

The fact that physicists have even conceived of such curious arguments as the locality loophole and the detection loophole shows that quantum correlations are of concern. It would be easy to present other 'experimentally verified' physical phenomena for which, in reality, much less precise and numerous data has been obtained. But physicists are human, and it is natural that they study with more determination that which is less easy to accept.

On the subject of the experiments that have demonstrated non-locality *à la* Einstein–Podolski–Rosen, and through that, non-separability *à la* Schrödinger, Shimony forged the term *experimental metaphysics*[71]. Is it an exaggeration? The reader can be the judge. It is, however, undeniable that in contemplating quantum interference for one and two particles, matter appears less ordinary than we thought it was. In revealing quantum physics, we could say that nature wanted to avenge itself on the positivism of the nineteenth century – experimental science, supposed to resolve all our doubts, plunges us back into surprise. It is time to set out for the overview of the interpretations.

·9·

Attempts at explanation

9.1 To the source of the surprise

During the lectures to the students in Fribourg, as with other people foreign to the world of physics, I have stated that my audience was driven to raise the same questions as the physicists: can we accept randomness in physical phenomena? Is interference determined by the mechanisms, and correlations at a distance by an exchange of information? In the presence of non-separability of physical properties, can we still talk about distinct entities?

We feel the need to interpret quantum phenomena because *we are not at ease with the indistinguishability principle and with some of its consequences.* The unease comes from the fact that this principle, which appears to be omnipresent in the microscopic world, is unknown in everyday life and even seems to contradict our normal perceptions – the reader must recall the game with the sets of cars in Chapter 2. Seen from this point of view, the interpretations can be divided into three broad categories:

- Those that accept the indistinguishability principle as a first principle – it is a principle like any other, formulated well enough and abundantly verified, therefore it is not unreasonable to set it down as a foundation. This is the orthodox approach, the one that I took in this book. Such a position is faced with the problem of explaining why the indistinguishability principle is not manifest in everyday life. The interpretations in this category differ basically on the answer to this question, often called a *measurement problem* (an answer to which I have not committed here).

- Those that endeavour to deduce the indistinguishability principle from notions considered more fundamental, but not of a physical nature. Such approaches are also 'orthodox' and seem *a priori* very promising. In practice, however, all the attempts made to date end up replacing the indistinguishability principle by something just as abstract (if not more so).
- Those that attempt to deduce the indistinguishability principle from physical notions that seem more compatible with the classical world of the everyday. Here is the realm of unorthodoxy. We have seen two failed attempts in this direction: the Heisenberg mechanism (Chapter 4); and the local-variable models (Chapter 7). Of this category, we will discuss the only satisfactory approach to date – the pilot wave interpretation of de Broglie and Bohm.

Interpretational issues have, of course, caused lengthy discussions, and it is impossible to do justice to everyone here. The reader who wants to delve deeper into these themes has an abundance of literature at their disposal[72].

9.2 The 'orthodox' approach

9.2.1 A satisfactory approach

As I have just said, the orthodox approach to quantum physics consists of accepting the indistinguishability principle[73]. To some people, this sounds like laziness: Einstein for instance, who believed that the task of physics was to unveil the *mechanisms* underlying phenomena[74]. Indeed, to people trained in classical physics, the acceptance of the weirdness of the quantum description may look like a failure. However, try and look at it as follows:

(1) Quantum physics fulfils all the criteria in order for us to be able to talk about an appropriate physical description,

namely: (i) we have a class of phenomena that is vast, moreover, which can be studied using an experimental method; (ii) we have a well-structured mathematical model; (iii) we have rules of correspondence between the objects of the theory and the data of the experiment; and (iv) in applying these rules of correspondence, we note excellent agreement between the predictions of the theory and the observations.

(2) Physics does not have to occupy itself with 'what things are' but with 'how they are connected to each other'. Now, the connections, the relationships, are precise, even at a fundamental level – the *connection* between indistinguishability and interference is well established. Why such connections? It is not the job of the physicist to answer this question.

In this light, the orthodox approach looks much more satisfactory. In short, the orthodox approach to quantum physics consists of accepting the idea that *physics does not describe the mechanisms but the relationships* – more precisely, the relationships that can been modified in part in order to be verified by the experiment[75]. Now, since all relationships predicted within the framework of quantum physics have been successfully verified, the physicist is content. The atomic scientist's dream of basing everything on an 'intuitive' mechanics of atoms proves impossible, but hope remains of basing everything on relationships that are *less intuitive, but well established,* between the constituents of matter.

However, the orthodox interpretation has its bogeyman, and this is . . . the everyday world! As I have stressed, quantum physics describes a *vast class* of physical phenomena – but it seems at odds with the physics of everyday objects. We have emphasized this difference at length in the previous chapters, and mentioned in Chapter 3 that the very existence of a boundary between the classical and the quantum world is not clear. Let's look more deeply into this point.

9.2.2 Bohr's vision

For any first description, one usually resorts to *Bohr's vision*[76]. In this vision, 'small' objects are quantum, subjected to the weirdness of the indistinguishability principle (of complementarity, in Bohr's words); on the contrary, 'large' objects, and in particular measurement apparatuses, are classical. When we make a measurement, we learn something and consequently the properties are modified. Remember the examples of Chapters 1 and 2 – one cannot measure the path on which a particle travels without disturbing the result of subsequent measurements. Before the measurement, the particle *had* some given properties (for instance, being able to show interference while being delocalized among all possible paths); after the measurement, it has *lost* some properties and acquired new ones (it will no longer show interference, but it has become localized on one precise path). This sudden modification of the properties, due to the measurement, has been called *collapse*: the set of properties 'collapses' onto a new one. Now, if one considers the collapse to be a real phenomenon, troubles rapidly arise. Nowadays, most physicists consider that collapse is simply a wrong description, although – as in the case of the hypothesis of superluminal hidden communication – it has not yet been possible to rule out all collapse models[77]. Anyway, instead of entering into the passionate debate about the reality of collapse, I prefer to move on and find the weak point of Bohr's vision – which is almost easy to find: the strict distinction between 'small' objects being quantum and 'large' ones being classical is entirely arbitrary, all the more so because large objects are supposed to be a large collection of small ones.

Thus, Bohr's vision agrees with our experience but seems inconsistent. Is there a consistent way of dealing with this question? Indeed there is one, as first noticed by Everett[78].

9.2.3 Everett's vision

In the approach to orthodox quantum physics initiated by Everett, everything is quantum. Everett's approach is intrinsically hard to understand: if we find quantum behaviour weird for particles, how can we imagine a whole world in which everything (ourselves included) are supposed to behave that way? But Everett's approach is consistent within itself, whence its appeal. Here is how a measurement is described. Suppose that before measurement, one has a particle delocalized between locations A and B; no detector has fired, the physicist is waiting for something to happen. After the measurement, two possibilities arise: p_A and p_B, defined by 'the detector in path A (respectively B) has fired and the physicist has become conscious that the particle was measured on that path'. There is nothing strange in that. But now, Everett goes on, it is not true that either one or the other of p_A or p_B has been realized: rather, the whole universe now finds itself 'delocalized' between the two possibilities[79]! In Everett's vision, everything is quantum: the world simply goes on developing relationships as described by quantum physics.

Actually, several years before Everett, Schrödinger had reached the same conclusion with his paradox of the cat. A cat plays the role of measuring apparatus, which detects whether a photon has been emitted and has triggered the explosion of a bomb, or not. The possibilities p_A and p_B are then 'photon not emitted, cat alive' and 'photon emitted, cat dead'. As we discussed in Chapter 7, Schrödinger came up with this example to point out how such a description is absurd – and it really does seem absurd, but it is consistent.

As happened for the topic of collapse, the viability of Everett's vision has been discussed at length, so I leave it to the interested reader to learn more about it. Some physicists have gone further than Everett: instead of saying that *the* universe is delocalized between p_A and p_B, they say that both p_A and p_B happen in *different* universes! In other words, each time an interaction takes place, the universe splits into as many copies as there were possible alternatives. In my opinion, this development goes beyond an orthodox approach to quantum physics, because it introduces unobservable elements, namely all the universes in which I am not conscious of existing[80]. This is usually called the 'many-worlds interpretation' – although sometimes, it is Everett's vision itself that is called that.

With Bohr and Everett's visions, we know the main points of the orthodox interpretations of quantum mechanics. We can turn now to other approaches.

9.3 Other foundations

Second on my list was the approach of those who attempt to derive the indistinguishability criterion from other principles that are judged more fundamental, but that are not of a physical nature. One example of this approach comes from the school called 'quantum logic'. The reader glimpsed the subject of quantum logic in Chapter 2, when we saw that the properties of a quantum system, unlike the properties of the sets of cars, are not connected to each other according to the rules of set theory. Let's take for example the work of the school in Geneva, a quantum logic approach initiated by Josef Jauch and continued by Constantin Piron.

Piron showed that one can *derive* the indistinguishability criterion from five axioms. The first three axioms are a formalization of the two following postulates. (I) If a physical system acquires a new property, it inevitably loses another that it possessed beforehand. An ordinary example: if I acquire the property 'being seated', I lose the property 'being standing' that I possessed beforehand. For a quantum example, we have already seen that the property 'exhibiting interference' can be lost to acquire the property 'being in a given path'. (II & III) Every property is the opposite of another. This simply means that if 'being seated' is a property, 'not being seated' is also a property. Each of us accepts such postulates much more easily than the indistinguishability criterion – it would be nice if we could derive the criterion only from postulates as intuitive as these. Unfortunately, things take a turn for the worse with axioms IV and V, which are strictly mathematical requirements[81] for which, despite significant efforts, neither Piron nor any member of his school knew how to find a simple interpretation. At this stage, then, we are faced with a choice: either we accept *all five* of Piron's axioms, in which case the indistinguishability principle is no longer a first principle but a consequence; or we admire the truly remarkable effort of the Geneva school, but we retain the indistinguishability criterion as a first principle – as I did in this book.

The school of quantum logic is only one example of a much broader class of interpretations that rapidly sink into deep epistemological discourse. All of these interpretations are not incompatible with the orthodox approach, and concede that if we restrict ourselves to the framework of physics we cannot say much. The program of looking resolutely outside physics to solve the conundrum of quantum phenomena is, in principle, very sound, but in my opinion has never been carried through satisfactorily – the surprising or 'incomprehensible' side of the indistinguishability principle does not disappear, is it simply pushed a degree further away, whether in the epistemological hypotheses or in the axioms.

9.4 The mechanistic interpretation of pilot waves

Among the unorthodox interpretations, I will focus on the most complete and successful: the interpretation of the 'pilot wave' initiated by Louis De Broglie and re-elaborated by David Bohm.

We saw in the first chapters of this book that quantum particles sometimes behave like corpuscles (each particle only stimulating one detector), sometimes like waves (interference). De Broglie's ingenious idea consists of exploring the possibility that *the corpuscle and the wave are both a physical reality*. More precisely, quantum particles could be corpuscles, very localized, which move around guided by a wave. It is the *wave* that explores all the possible paths, and it is the modification of the properties of the wave that influences the 'choice' made by the corpuscle at each beam splitter. It is just like a cork floating in a river, downstream of an island: certainly, the cork passed on only one side of the island; nevertheless, its trajectory after the island is also influenced by the water that has taken the other path. This example illustrates the explanation of Young's double-slit experiment by a pilot wave.

Naturally, things are not that simple, otherwise every physicist would accept this interpretation, quantum physics would be a form of fluid physics and this book would never have been written. First, unlike water, the hypothetical quantum wave does not have to transport energy. In fact, the quantum wave that guides the corpuscle should be unobservable – once again, we are in the presence of an interpretation that introduces an unobservable element into physics. Secondly, the pilot wave is not a wave in three-dimensional space, like the waves on the ocean or sound waves. To understand this, it is enough to recall that the interference does not depend only on the difference *in length* of the paths, but on any difference at all (the spin in the Rauch experiment, the state of the energy in the Constance experiment, the polarization in the Aspect experiment...). Therefore, the pilot wave must be sensitive to every modification, if we want it to be that which manages the interference. Thirdly, while the pilot wave provides a hidden variable description of one-particle interference, one cannot beat Bell's theorem. In order to explain quantum correlations at a distance, it is necessary to postulate that the operations carried out on one particle change the wave affected by the other particle instantaneously[82] – in other words, the pilot wave is a hidden *non-local* variable.

In the pilot-wave interpretation, there is nothing ridiculous, nor anything that openly opposes other theories in physics. Notably, the instantaneous modifications of the wave do not contradict relativity since they are unobservable and they could not be exploited to send a signal faster than the speed of light. John Bell took the pilot-wave theory seriously. Einstein himself, according to his biographer Abraham Pais, held a lot of hope for the work of De Broglie on the pilot wave, but he published nothing himself on the subject – a silence that is certainly significant[83].

9.5 Further notes for a balance

9.5.1 Randomness and determinism

The interpretation of quantum physics is an area of disagreement among physicists[84]. This was one of the points that I stressed to the students at Saint-Michel in Fribourg. One of the students reminds me of the issue of randomness, which I had left unresolved. Now we know that randomness means different things according to the different interpretations. In Bohr's vision, the outcome of a measurement is *objectively* random[85]: this means that before the measurement, the outcome was intrinsically undetermined, even a being who would have access to the state of the whole universe could not have predicted it. In Everett's vision, there is actually no randomness in a single measurement, only relationships build up; if one repeats an experiment many times, it turns out that in almost all cases (all 'worlds') the sequence of outcomes will respect the probabilities predicted by quantum theory. In Bohm's interpretation, the randomness is subjective: it arises because we don't have access to the pilot wave, just like the randomness of coin tossing arises from our lack of control over all the meaningful parameters. The students are quite grateful now that I had not suffocated them with all this complexity at the beginning of my first lecture . . . And there is a last point I want to make.

Some people say that quantum physics has banished determinism from physics, replacing it with randomness. This claim is 'reasonable', but not rigorous. Suppose you believe in strict determinism (I don't), according to which all the details of the universe's story have been set in stone, and are now simply unfolding in time. Then, in particular, the fact that you 'choose' to do a quantum measurement was pre-determined, and the result of the measurement was also pre-determined. Also, as we mentioned, within such a view of the world, Bell's theorem becomes meaningless: the written history of the world is a huge non-local hidden variable that explains everything. Of course, this is the least of the inconveniences (the loss of any form of human freedom is a significantly more dramatic consequence). But it shows that quantum physics cannot ultimately answer for our worldview. What is true is the following: if you accept some reasonable freedom for your choices – so that, in particular, you can choose to prepare a given quantum state and to make a given measurement on it – then the outcome of the measurement is normally out of your control: if you accept some 'indeterminism' for man, then you must also accept a form of 'indeterminism' in Nature[86].

9.5.2 My position

'And you, what do you think of all that?' A most natural question, indeed! Well, I think that it is important to know the basics of the interpretational debates, but that it is also safe not to devote one's life to them. Note that, apart from a few 'prophets', physicists have come to realize the crucial role of entanglement only in the last, say, twenty years. Note also that quantum physics is seriously at odds with gravity (general relativity, the concepts of space-time, etc.): at present, nobody knows how to reconcile both theories, because in all the phenomena that we usually observe, one of the two theories plays basically no role so that we can make predictions with the other one – but the tension exists[87]. To be concrete: if quantum objects would 'perceive' a different space-time than the one to which we are accustomed, then it may be that interference and entanglement are no longer astonishing features, but rather the most natural ones! In short, as long as quantum physics and gravity have not been reconciled, the final word on interpretations will be elusive.

That being said, I think two challenges are worth undertaking. The first challenge consists of trying to *give entanglement the priority over indistinguishability*. As the reader has noticed, in my opinion entanglement is really the heart of quantum mechanics: if there were no entanglement, but only one-particle interferences, then I would unhesitatingly accept a pilot-wave theory, certainly the most economical one in terms of concepts involved (it just uses the concepts of classical physics). To put it differently, I know that a photon is a quantum object because I know that two photons can become entangled[88]. In this book I had to follow the usual path (from one to two particles) because the other path, which starts from entanglement, does not exist yet.

The second challenge consists of trying to *experimentally test the difference between Bohr and Everett's visions*: is there a definite boundary between the quantum and classical worlds (the so-called

Heisenberg cut), or not? We have seen in Chapter 3 that the boundary is not reached for large molecules. Actually, if the boundary exists, it is certainly not only a matter of 'size': while clusters of very few metallic atoms quickly show classical behaviour, cold diluted gases containing tens of thousands of atoms can still show entanglement! The search for the boundary is a profound issue, ultimately related to the status of irreversibility in physics[89]. What shall we find? I tend to be more inclined to Bohr's vision[90], and thus hope that a boundary will be found – its detailed exploration will then provide work for physicists for many years.

·10·
In my end is my beginning

10.1 Variations

Nature makes innumerable variations on the theme 'indistinguishability principle'. In certain variations the theme is obvious, in others it is more hidden – exactly as in music. I have presented the phenomena that most directly show the specificity of quantum physics; but to think that this book contains all of quantum physics would be like killing Schubert's *Trout* in the first octaves.

Think of the historical successes of quantum physics – the comprehension of chemistry, the discovery of the atom first and the world of elementary particles afterwards. Think of the research domains that, apart from increasing our understanding of Nature, have produced important applications – quantum optics with the laser, solid-state physics with semiconductors and superconductors, atomic physics with the atomic clocks required for precise synchronicity (for instance in the GPS system). Think of the most recent but fascinating domains such as mesoscopic physics or the physics of the Bose–Einstein condensates[91] produced for the first time in 1995. Think of the frontiers of cosmology, for instance the so-called inflation that is supposed to describe the very first instants of our universe. Think of the impetus that the needs of physics have given to certain branches of mathematics. Think of all that, and you will have a more comprehensive summary of what quantum physics is – each of these points deserves many books for itself alone.

But our itinerary stops here . . . almost. One of the most intriguing quantum phenomena is so-called *teleportation*. Teleportation has attracted a great deal of interest recently. Now, this phenomenon involves *three* particles: it is thus the natural continuation of the single-particle and two-particle phenomena on which this book was based. Its full explanation would require one more conceptual tool than those introduced so far. I will satisfy myself with a description of what teleportation is, through one last window on the quantum world.

10.2 Quantum teleportation

First of all, be reassured (or disappointed): quantum physics does not allow us to teleport matter[92]. What is going to be teleported is information, namely *the properties* (the state) of a quantum particle. Specifically, these properties 'disappear' from particle A that initially carried them, and reappear on an arbitrarily distant particle C. This accounts for two particles; the third one, particle B, plays a crucial role that we are going to describe. Before going into the protocol of teleportation in more detail, we must grasp the difference between ordinary transport and teleportation of information. In both processes, information is moved from its initial location to a distant one. Usually, however, during the transport, the information is available at any intermediate location. In teleportation, this is not the case: the information disappears in one location and reappears in another, without having been available in any intermediate location.

The teleportation protocol is illustrated with spins in **Fig. 10.1** In the initial step, particle A is prepared with its spin pointing in a given direction; particles B and C are prepared in an entangled state. Remember that in entanglement, none of the individual spins has a well-defined direction, but the relation between the directions is well defined – say, the two spins of B and C point in the same direction[93]. Particles A and B are then brought together, while particle C remains separate. Now comes the quantum trick: particles A and B are measured in a clever way, called *Bell measurement*: roughly, they are asked the question 'Are your spins identical

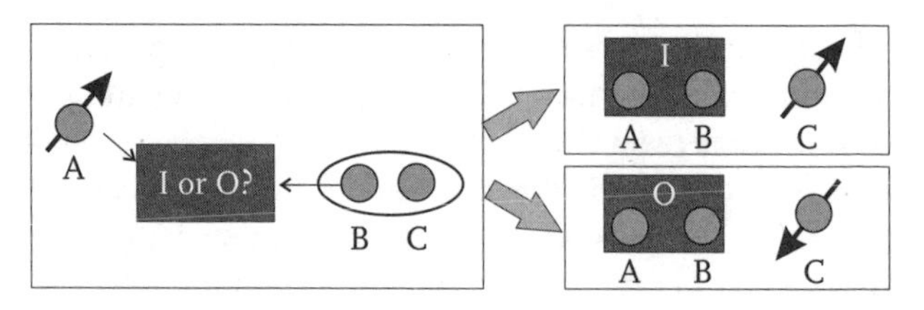

Fig. 10.1 Sketch of the teleportation protocol. See text for details; I and O stand, respectively, for identical and opposite.

or opposite?' This question must be asked in such a way that the particles can answer it directly, without having first to commit to a specific value for their individual spins. Suppose the answer is 'identical': then the spin of A is identical to the spin of B; but because of entanglement, the spin of B was identical to the spin of C – so now the spin that C acquires after the measurement of A and B is exactly the spin that was initially carried by A. Teleportation has succeeded. Suppose that the answer is 'opposite': a reasoning similar to the previous one shows that now the spin of C acquires the spin that is opposite to the one initially carried by A. Has teleportation failed? Not quite: one just has to inform the physicist who holds particle C that he has to reverse the spin in order to achieve teleportation. Thus, teleportation can be made successful in all cases.

Let's go back to a few points, starting from the end. First, the result of the measurement of A and B *must* be communicated to the physicist who holds C: this result is random, so without knowing it, one cannot know whether particle C carries identical or opposite spin to the desired one. But in turn, this implies that – like entanglement – teleportation cannot be used to send a signal faster than light: classical communication (internet, telephone) is required to finish the protocol. Secondly, we can justify the claim we made that the information on the spin is not available in any intermediate location between A and C. In fact, on the one hand, particle B never carried any information on the spin of A, and the measurement on A and B modifies at a distance the spin of C without any signal propagating towards C. On the other hand, the communication of the result of the measurement carries only one bit of information ('identical' or 'opposite', 'yes' or 'no') while the spin of A may have pointed in any arbitrary direction (infinitely many possibilities). Finally, let me say that things are a bit more complex: specifically, the measurement of A and B can in reality give four outcomes (not just two). But the idea is the same, and in all four cases one can achieve perfect teleportation by sending the corresponding information.

Teleportation has been experimentally demonstrated several times to date, in different configurations, mostly with photons[94] but also with massive particles[95] (in this last case, the particles were not very far apart from one another, but the effect is still remarkable).

10.3 Epilogue

We have chosen a very precise path through the forest of quantum phenomena, the path of surprise and wonder. Aristotle recommended this path in order to approach Nature. But most of the phenomena that surprised Aristotle no longer surprise us. Is it that quantum physics surprises us simply because it is 'new'? Will humanity become accustomed to quantum physics once it is better known, just as we have become accustomed to heliocentrism although our primary sensations don't suggest it? I rather believe that more surprises are yet to come. That is why I like to conclude my presentations of quantum physics with a formula, which conveys both the beauty and the dangers of the path to knowledge[96]:

Not farewell,
But fare forward, voyagers!

Endnotes

1. R. Feynman, R. Leighton, M. Sands, *The Feynman Lectures on Physics: Quantum Mechanics* (Addison-Wesley, Redwood City, 1989).
2. R. Feynman, *QED: The Strange Theory of Light and Matter* (Princeton University Press, Princeton, 1985).
3. The facts comprising this anecdote have been relayed to me by Alain Aspect. Of course, more than twenty years later, the reproduction of the words and thoughts are sheer guesses I have made for stylistic reasons.
4. I have borrowed this expression from a text by Alain Aspect, 'John Bell and the second revolution', written as the introduction to: J.S. Bell, *Speakable and Unspeakable in Quantum Mechanics* (Cambridge University Press, 2nd edn, 2004).
5. The experiment will be published in 1986: P. Grangier, G. Roger, A. Aspect, *Europhys. Lett.* **1**, 173 (1986).
6. For the experienced reader, I want to stress that there is an important difference between the 'first' and the 'second quantum revolution'. The first one, which took place early in the twentieth century, marks a sharp break in physics: classical physics is found in utter disagreement with observed phenomena and had to be replaced by something else. The second quantum revolution does not break away from the first one, the predictions are the same and no phenomenon has been observed that departs from them. The second revolution deals rather with the way we look at quantum physics, and is based mainly on the awareness that (1) quantum physics applies to *individual* systems, not only to ensembles, although its predictions can be verified only for ensembles, and (2) *entanglement* between different subsystems plays a crucial role. In summary, the quantum physics I am speaking about has remained the same since 1926, but some elements that were considered as fundamental issues some time ago (Heisenberg

uncertainty relations, wave mechanics, Schrödinger equation . . .) are now perceived as derived notions – hence they will play almost no role in this text. Obviously, nobody is claiming that the second quantum revolution is the final word on quantum physics.

7. 'For it is owing to their *wonder* that men both now begin and at first began to philosophize'. Aristotle, *Metaphysics*, Book 1 Part 2 (translation by W.D. Ross, The Internet Classics Archive, URL classics.mit.edu/Aristotle/metaphysics.1.i.html); my italics.
8. If the reader has skipped the Prologue, as I often do myself when reading a book, they had better step back and read it now – here, and in a few more places later, I refer explicitly to its content.
9. Throughout the book, I will use the word 'particle' to describe quantum objects in general (from the so-called elementary particles, up to atoms and molecules). This is the usual wording, but because it is equivocal and misleading – we shall soon learn that these 'particles' have a highly 'non-particle-like' behaviour – some authors have attempted other names, such as 'quantons' in: J.M. Levy-Leblond, F. Balibar, *Quantics: Rudiments of Quantum Physics* (Elsevier, Amsterdam, 1990).
10. Actually, this cycle can go on many times, but not infinitely many. If the difference in the length of the paths becomes larger than a property of the particle called the *coherence length*, then the interference vanishes.

    For the experts: a quantitative understanding of interferometry requires the description of the particle as a wave packet. In such a description, the length $2L$ of the main text is just the (central) wavelength of the packet, while the coherence length is the extension of the wave packet.

    At this stage, one may raise doubts about the didactic approach to quantum physics presented here. A first doubt would be, shouldn't I have introduced the wave-particle duality from the very beginning? The answer to that is a decided No: waves will be indeed discussed in Chapter 2, but from Chapter 6 on it will become clear that wave-particle duality fails to describe all of quantum physics (because it fails to describe entanglement). A second doubt would be, shouldn't I have chosen other forms of interference instead of path interference? I have no clear-cut answer to that: on the one hand, I feel that path interference is more striking for an initial presentation; on the other hand, as soon as I must give a slightly more elaborate discussion,

I usually resort to polarization. An approach based on polarization has the default that one must first understand what polarization is, while 'being in a path' is a property associated to common perceptions. For a very good presentation of quantum mechanics based on polarization, I suggest: G.C. Ghirardi, *Sneaking a Look at God's Cards* (Princeton University Press, Princeton, 2003).

11. The synthetic step is necessary to science, as recognized for instance by Kant, one of whose main concerns was establishing the validity of synthetic reasoning – in my opinion, he failed at the task, but this is a minor point here.
12. Unfortunately, indistinguishability is an equivocal word within quantum physics! One meaning is the one conveyed here, namely a criterion for observing interferences. The other, more traditional, meaning is related to the fact that two particles (say, two electrons) are identical: there is no way of telling which is which after they have interacted. Thus, one speaks of indistinguishable particles. Both meanings are widely used nowadays, so I did not think it wise to remove the ambiguity. In this book, only the first meaning will be used.
13. T.S. Eliot, *Four Quartets* (Faber and Faber, London, 1959): Burnt Norton, vv. 42–43.
14. If the reader has some energy left, here is an intriguing consequence of quantum interference: the possibility of detecting the presence of an object without any particle touching it (an effect called *interaction-free measurement*). The idea is quite simple: let's consider the apparatus in **Fig. 1.3**, the balanced Mach–Zehnder interferometer such that all of the particles take the output *RT or TR*. Let's suppose that we introduce an obstacle that blocks one of the paths: as a result, there are no longer two indistinguishable paths that interfere, and (in accordance with the apparatus in **Fig. 1.1**) half the particles take the output *TT or RR*. Now, each time a particle is detected at this output, we know (i) that there was an obstacle on the path; and (ii) that the particle has taken the other path. Therefore the particle informs us of the presence of an obstacle with which it has never interacted! A slightly more sophisticated version of this idea, based on what is called the quantum Zeno effect, can be found in: P. Kwiat, H. Weinfurter, A. Zeilinger, *Scientific American*, November 1996, p. 52.

15. For experts: the 'set' is nothing but the phase space, or configuration space, which is the space of states of classical physics. A point in the set is a pure state, that is, a state of maximal knowledge: all of the properties are perfectly determined; a subset is a mixed state, representing a statistical ensemble. Note in particular that the phase space has the structure of a set, but not that of a vector space: by 'adding' two points of the phase space, one finds another possible state, but completely unrelated to the original ones (all pure states are orthogonal in classical physics). Mathematically, this is the great difference between the quantum and classical descriptions of physical systems. This is all clear . . . once someone has told you! I learnt it from François Reuse, who introduced me to the principles developed by the Geneva school. A formalization of these ideas may be found in: C. Piron, *Foundations of Quantum Physics*, (W. A. Benjamin, Reading, 1976).
16. This statement is precise and requires source emphasis. The point is that one can ask questions about any system's 'essence' or about its 'accidents'. For instance, an electron is a particle with a given mass, charge, spin: these values constitute the 'essence' of what an electron is, if a particle has different values then it is not an electron. Then come the 'accidents': the electron may be here or there or even delocalized; its spin might point in one or other direction; and so on. We call the *state of the system* a determination of the accidents – obviously, the fact of being there rather than here does not change the fact that the system in question is an electron. In quantum mechanics, it is questions about the state (the 'value' that each accident takes) that do not combine according to the rules of set theory; but the questions about the essence do – in technical terms, one cannot define a superposition between 'being an electron' and 'being a proton'.
17. There are so many books that present the birth of quantum physics out of classical physics that I don't dare suggest any of them as being the best!
18. For a comprehensive history of atomism, refer to: B. Pullman, *The Atom in the History of Human Thought* (Oxford University Press, Oxford, 1998).
19. Based on the same principle, one quickly realizes that there exist in principle an infinite number of interferometric apparatuses: each

time we arrange with mirrors, slits, lenses, etc. to have two indistinguishable paths, we are in the presence of an interferometer. In this book, we shall use only Mach–Zehnder or Young's as experiments involving a single particle; two other interesting interferometers that also work for classical light are the one invented by Sagnac, and the famous one invented by Michelson and Morley to measure the speed of light relative to the hypothetical ether. Both are described in any advanced book of optics.

20. An account has been written by one of the protagonists: C.H. Townes, *How the Laser Happened. Adventures of a Scientist* (Oxford University Press, Oxford, 1999).
21. An experimental (the first particle accelerator) is very well described at a popular level in: B. Cathcart, *The Fly in the Cathedral* (Penguin, London, 2005).
22. C. Magris, *Danube* (Harvill Press, London, 1999).
23. One of the first articles reporting on these experiments was: H. Rauch, A. Zeilinger, G. Badurek, A. Wilfing, W. Bauspiess, U. Bonse, *Phys. Lett.* **54A**, 425 (1975). For a comprehensive view of neutron interferometry, refer to: D. Greenberger, *Rev. Mod. Phys.* **55**, 875 (1983).
24. Lovers of topology or art will have noticed that the spin goes around on a Möbius strip, like the ants in the famous Escher engraving.
25. Experts should not fear heterodoxy here: I am not contesting that quantum theory predicts only statistical quantities, which can only be measured by detecting many identically prepared systems. The remarkable fact is that the interference patterns build up even when particles are sent to the detectors one after the other.
26. Experts recognize here the problematic status of the theory of decoherence. Decoherence is an important effect – a physical system is never completely isolated, it interacts with an environment. Through this interaction, the number of degrees of freedom involved in the full description of the evolution increases rapidly. Roughly speaking, one can say that the information about the state, initially concentrated on the system under study, is rapidly diluted into the environment. Mathematically, it can be shown that such an interaction indeed happens very rapidly for large objects, and has the effect of destroying interferences. However, this does not solve the conceptual problem we have raised, which can now be rephrased as follows: is decoherence everything in the transition from quantum to classical, or is

some other physics going to appear on some level? For a remarkably clear account of the achievements and the limitations of the decoherence program, refer to: M. Schlosshauer, *Rev. Mod. Phys.* **76**, 1267 (2004). In this book, more will be found in Chapter 9, when discussing the difference between Bohr and Everett's theories of measurement, and the issue of irreversibility.

27. Other beautiful experiments are those performed by Serge Haroche's group in Paris, for instance: M. Brune, E. Hagley, J. Dreyer, X. Maître, A. Maali, C. Wunderlich, J.M. Raimond and S. Haroche, *Phys. Rev. Lett.* **77**, 4887 (1996); P. Bertet, S. Osnaghi, A. Rauschenbeutel, G. Nogues, A. Auffeves, M. Brune, J.M. Raimond, S. Haroche, *Nature* **411**, 166 (2001). A detailed description of these experiments requires a good knowledge of physics.
28. Buckminster Fuller is known to have been somewhat extravagant. In his domes, he saw not just a way of covering large areas with a relatively light structure, but also a form inspired by nature and by the modern vision of a world closed in over itself. He died two years before the discovery of C_{60}.
29. The first experiment of Zeilinger's group on the interference of C_{60} molecules is: M. Arndt, O. Nairz, J. Vos-Andreae, C. Keller, G. van der Zouw, A. Zeilinger, *Nature* **401**, 680 (1999). Along the way, other experiments were performed by the same group, either with different apparatus or with C_{70} molecules (which look more like rugby balls than footballs), then with even larger ones.
30. I am indebted for this remark to Marek Zukowski, a theoretician from Gdansk who collaborates with Zeilinger's group.
31. Other large systems (actually much larger) showing collective quantum behaviour are cold atomic gases, as described, e.g., in: B. Julsgaard, A. Kozhekin, E.S. Polzik, *Nature* **413**, 400 (2001). This experiment even demonstrates *entanglement* between two such systems.
32. Contrary to widespread belief in the physics community, the Heisenberg mechanism is *not* the way Heisenberg first derived his famous 'uncertainty relations' in 1925. He invented it some years later, when trying to find a more illustrative way of presenting his discovery. It first appeared in: W. Heisenberg, *Physical Principles of the Quantum Theory* (Dover Publications, New York, 1930).

33. See, for instance, paragraph 1.8 of Feynman's lectures, or Complement D_I of the textbook: C. Cohen-Tannoudji, B. Diu, F. Laloë, *Quantum Mechanics* (Longman Scientific & Technical, Essex, 1977). Some detail for the experts: as stated in the text, the Heisenberg mechanism does exist and is an effective source of decoherence; in particular, the calculation leading to $\delta x \delta p \sim \hbar$ is correct for the measurement of the position of a particle with a photon. But there are two points to be stressed. First, it is not a universal source of decoherence: in the Constance experiment for instance, for the electron it holds something like $\delta x \delta p_e \sim \hbar$; but for the atom, $\delta x \delta p \sim 0$; still, interference disappears. Secondly, it is *wrong* to identify $\delta x \delta p \sim \hbar$ with the 'uncertainty relation' $\Delta x \Delta p \geq \hbar/2$. The first relation means that a measurement of the position with precision δx will entail a modification of the state measured by a perturbation in the momentum δp. The physical setup is one of an intermediate measurement (here, of the position), that modifies the state; a setup that we have already met in Chapter 1, then more explicitly in Chapter 5 on quantum cryptography. The second relation refers to intrinsic variances: if one prepares a state that is well localized in space, then it will be delocalized in momenta, and vice versa. To verify the uncertainty principle, one must take a source that prepares many particles in the same state, measure x on half of the particles and p on the other half, then compare the variances; certainly not measure first x and then p on each particle!
34. As the reader has guessed, it was published: S. Dürr, T. Nonn, G. Rempe, *Nature* **395**, 33 (1998).
35. Experts will immediately notice that, as such, this is not a very practical scheme: the two paths should be equal to within a fraction of the wavelength. For instance, using light in optical fibres, the wavelengths are of the order of one micrometre, whereas Alice and Bob may be separated by tens of kilometers. There are, however, simple ways out of this problem: the two channels are two modes supported by the same optical fibre, for instance two orthogonal polarization modes.
36. A review article exists covering all of the results discussed in this chapter: N. Gisin, G. Ribordy, W. Tittel, H. Zbinden, *Rev. Mod. Phys.* **74**, 145 (2002). But the explosion did not finish there: at the moment of my writing (less than three years later), this review, while

very good in exposing the basic issues, is already almost obsolete for experts working in the field.

37. It is useful to draw a comparison between quantum cryptography and other applications of quantum physics. Consider this example: the laser is without doubt an application of quantum physics; however, control of the coherence of individual quantum systems is not required to build it – stimulated and spontaneous emission can be derived from the atomic hypothesis and purely thermodynamical considerations, as Einstein did back in 1917. In other words, a laser is based on quantum physics, in the same way as, for instance, one needs quantum physics to understand why some materials are electric conductors and some are not. On the contrary, quantum information provides the *first applications for which the coherence of individual systems is strictly required*: one can do quantum cryptography by sending one photon after the other, but cannot do it by sending classical pulses of light, even though this light is also ultimately 'made' of photons.
38. J.D. Franson, *Phys. Rev. Lett.* **62**, 2205 (1989).
39. For the experts, a few more words can be said. The source is a special kind of crystal in which the phenomenon of parametric down-conversion can take place. This means that when it is irradiated with a laser (the 'pump') at the appropriate frequency, the crystal is basically transparent but sometimes absorbs one photon of the pump and creates two new photons. This process is such as to conserve both energy and momentum, which means specifically that no 'record' is left in the crystal that the down-conversion has taken place. The two new photons are emitted 'at the same time', but this time can be anywhere within the coherence time of the pump (since this is the 'size' of each single photon in the pump).
40. This may sound at odds with the content of Part 1: isn't it true that each particle meets two indistinguishable paths, since the time of emission in intrinsically unknown? The answer is, no, the paths of each particle are distinguishable in principle. Here is the proof: let's consider the particle that leaves on the left. At the moment of leaving the source, it does not carry with it any information about what it will encounter on its path, because the physicist is free to modify the apparatus after the particles are emitted, *a fortiori*, it does not carry with it any information about what the *other* particle, that which flew away on the right, will encounter. It is possible then that the physicist

who is on the right has removed the interferometer and has replaced it with a simple detector. Since the particles have been emitted at the same time, the knowledge of the time of arrival of the particle on the right automatically reveals the time of emission of the pair. The apparent indistinguishability of paths for the particle on the left can thus be destroyed by communicating with the physicist who is on the right. In conclusion, the two paths (short and long) of one side of the apparatus *could be distinguishable,* according to what the physicist on the other side does. We see here that being one of a pair considerably modifies the analysis of the distinguishability for one particle. For experts: for the Franson interferometer to exhibit two-particle interference, the difference between L and S must be larger than the coherence length of each single photon and shorter than the coherence length of the pair (that is, of the pump laser).

41. The first popular science article I ever read on many-particle correlations is still one of the best: D. Greenberger, M. Horne, A. Zeilinger, *Physics Today* 46, August 1993. For a comprehensive view of the topic, I recommend experts read part II of: A. Peres, *Quantum Theory: Concepts and Methods* (Kluwer, Dordrecht, 1998).
42. The hypothesis of unobservable supraluminal communication has been studied in a few works, and quantum physics emerges strengthened from this confrontation. Meaningful titles are: V. Scarani, W. Tittel, H. Zbinden, N. Gisin, *Phys. Lett. A* **276**, 1 (2000); A. Stefanov, H. Zbinden, N. Gisin, A. Suarez, *Phys. Rev. Lett.* **88**, 120404 (2002); V. Scarani, N. Gisin, preprint quant-ph/0410025 at the server xxx.lanl.gov.
43. Believe it or not, there are even some patents that claim that one can use quantum correlations to signal faster than light – some of these even give the speed . . . Patent agents systematically reject proposals of *perpetua mobilia,* which would violate the first or second law of thermodynamics, but seem more open to accept a violation of special relativity.
44. At least no ordinary signal: recall the discussion about a hypothetical hidden communication. For simplicity, we shall not mention this heterodox assumption from now on.
45. Historically, these hypothetical parameters were called *local hidden variables,* often only *hidden variables,* because quantum theory does not use them explicitly. But their being 'hidden' or not turns out to be irrelevant: Bell's theorem depends only on the fact that these

variables are 'local' (established at the source and independent of what is going to happen later).

46. Indeed, on the one hand we want Alice and Bob to be able to choose their measurement, and in order to choose, there must be at least two alternatives. On the other hand, a measurement that always has the same outcome would be trivial, therefore we must accept at least two possible outcomes for each measurement.
47. This is not a restriction – let me explain why. In the spirit of the Bell theorem, we could think that the result of each measurement is determined by (i) the parameters established at the source, (ii) the choice of the experimenter, and (iii) some parameters present in the measurement apparatus, which may be beyond our control. Only this last point is new in relation to what was discussed in the text, and I am going to show that it brings nothing that could invalidate the Bell theorem.

   Let's consider the particle that arrives at Alice's end, the same reasoning being just as valid for the one that arrives at Bob's end. This particle carries with it the parameters of point (i) that depend neither on Alice's nor Bob's choice of measurement, because the choices can be modified after the particle has left the source. On arriving at the measurement apparatus, the particle discovers that Alice has chosen measurement A [point (ii)] and discovers new parameters, those of point (iii), which possibly depend on A. This can seem an obstacle to proving the Bell theorem, because the result *a*' of the measurement A' is not defined – we are lacking the parameters of point (iii) associated with A'. The Bell theorem remains nevertheless valid, for the following reason: it is not essential that *a* and *a*' can actually be determined, because we are not looking to explain how the particle that travels towards Alive 'chooses' its result. What is essential is that neither Alice's choice, nor the new parameters of point (iii) depend on the choice that *Bob* makes on his side! If this is the case, then *the correlations continue to be determined only by the parameters established at the source*, the additional parameters of point (iii) can only add an independent random element at Alice's end and at Bob's end, and therefore can only *diminish* the correlations.

   To summarize, the only method of bypassing the Bell theorem in a classical context consists of supposing that, in one way or another, the particle that arrives at Alice's end is informed of the choice that Bob has made, that is, that there is a signal that is propagated almost instantaneously. The other known method of bypassing the Bell theorem is in calculating with quantum theory. As I said in the text of Chapter 6, at the present time we do not know if these two methods coincide or not – the possibility of supraluminal communication hidden in a special reference system has not been excluded by the experimental method. In line with what we have learned until now, I prefer to think that there is no communication, and that the quantum correlations define a mode of correlation unknown in the classical world.

48. The original derivation by Bell can be found, together with many other stimulating papers by John Bell, in: J.S. Bell, *Speakable and Unspeakable in Quantum Mechanics* (Cambridge University Press, Cambridge, 1987; 2nd edn 2004).
49. J.F. Clauser, M.A. Horne, A. Shimony, R.A. Holt, *Phys. Rev. Lett.* **23**, 880 (1969)
50. Mermin's articles for the general public are collected in: N.D. Mermin, *Boojums All The Way Through*, (Cambridge University Press, Cambridge, 1990).
51. D.M. Greenberger, M. Horne, A. Zeilinger, in: E. Kafatos (ed.), *Bell's Theorem, Quantum Theory, and Conceptions of the Universe* (Kluwer, Dordrecht, 1989), p.69; N.D. Mermin, *Am. J. Phys.* **58**, 731 (1990).
52. Realism is a philosophical method, according to which our senses are triggered by an existing world outside us, and our knowledge can reach some truth (although partial) about this world. The great representatives of realism have been Aristotle and Thomas Aquinas. In the history of philosophy, realism has been opposed, for instance, to rationalism (we know with certainty only a few truths evident to our mind, like Descartes' *cogito*, and the statements that can be logically derived from them) and to empiricism (we know only a bunch of sensations, but there is no truth in synthetic statements). Thus, 'local realism' is just the opposite of realism, because the tenets of the first position must reject the knowledge of nature we have gained through experiments – it would then be more correct to speak of 'local surrealism', as I heard once from Hans Briegel.
53. A splendid account of these debates is provided by Chapters 5 and 6 of: M. Jammer, *The Philosophy of Quantum Mechanics* (J. Wiley & S., New York, 1974).
54. A. Einstein, B. Podolski, N. Rosen, *Phys. Rev.* **47**, 777 (1935).
55. N. Bohr, *Phys. Rev.* **48**, 696 (1935).

56. E. Schrödinger, *Naturwissenschaften* **23**, 807 (1935), English translation in: J.A. Wheeler and W.H. Zurek (ed.), *Quantum Theory and Measurement* (Princeton University Press, Princeton, 1983).
57. In classical physics, the space of states of a composed system is the *Cartesian product* of the spaces of states of each constituent; properties are related to one another according to the rules of set theory, as it should be. In quantum physics, the space of states of a composed system is the *tensor product* of the spaces of state of each constituent, which is a vector space itself, as it should be.
58. With his argument of the cat, Schrödinger anticipated the approach to the theory of measurement proposed by Everett. There is more on this issue in Chapter 9.
59. I heard this story from Shimony himself, on one of his visits to Geneva.
60. See, for instance, the proceedings of a conference recently held in his memory: R.A. Bertlmann, A. Zeilinger (ed.), *Quantum [Un]speakables* (Springer Verlag, Berlin, 2002).
61. A. Aspect, P. Grangier, G. Roger, *Phys. Rev. Lett.* **47**, 460 (1981).
62. The Franson interferometer was proposed later, in 1989; it does not test polarization entanglement, but the sort of entanglement described in Chapter 6, which is called *energy-time entanglement* (because the particles are created at the same time in an energy-conserving process).
63. A. Aspect, J. Dalibard, G. Roger, *Phys. Rev. Lett.* **49**, 1804 (1982).
64. For the experts, a review article exists: W. Tittel, G. Weihs, *Quant. Inf. Comput.* **2**, 3 (2001).
65. G. Weihs, T. Jennewein, C. Simon, H. Weinfurter, A. Zeilinger, *Phys. Rev. Lett.* **81**, 5039 (1998).
66. W. Tittel, J. Brendel, H. Zbinden, N. Gisin, *Phys. Rev. Lett.* **81**, 3563 (1998).
67. I was witness to Gisin's surprise, who did not even know of the existence of this series of posters, when a colleague came to congratulate him.
68. Technically, one says that the *fair-sampling assumption* is respected in the first example but not in the second one.
69. Bell's wording is worth quoting: 'It is difficult for me to believe that quantum mechanics, working very well for currently practical set-ups, will nevertheless fail badly with improvements in counter

efficiencies' (*speaking*..., page 109). In a letter to me, Abner Shimony quoted a similar comment by Mermin: 'If local realism is to be saved by a failure of quantum mechanics to describe the entire ensemble of photon pairs, but in such a way that those pairs actually detected nevertheless do agree with the quantum theoretical predictions, then God is considerably less subtle and significantly more malicious than I can bring myself to believe'.

70. M.A. Rowe, D. Kielpinski, V. Meyer, C.A. Sackett, W.M. Itano, C. Monroe, D.J. Wineland, *Nature* **409**, 791 (2001).
71. A. Shimony, *Br. J. Philos. Sci.* **35**, 25 (1984), in Section 5.
72. Here are a few suggested readings. For an approach to interpretations that closely follows the historical evolution of quantum mechanics, a reference work is: M. Jammer, *The Philosophy of Quantum Mechanics* (J. Wiley & S., New York, 1974). Of course, Jammer's book was written before Rauch and Aspect's experiments – actually, before *any* of the experiments described in this book had been performed. A collection of radio interviews provides a good summary of the interpretations of quantum physics that are still the most widespread today, each presented by one of its principal supporters: P.C.W. Davies, J.R. Brown (ed.), *The Ghost in the Atom* (Cambridge University Press, Cambridge, 1986). One of the most recent essays focusing on interpretations is: J. Baggott, *Beyond Measure* (Oxford University Press, Oxford, 2004).
73. For experts: what needs to be assumed is the description of quantum systems with Hilbert spaces. All the rest basically follows: the probability rule is Gleason's theorem, the non-relativistic Schrödinger equation is obtained by requiring Galilean invariance, and so on. The foundations of relativistic quantum physics are much less strong, although the agreement between the predictions and the observations are astoundingly precise.
74. This is probably one of the sources of Einstein's strength, and explains why relativity is attributed to him rather than to Poincaré. In fact, both derived the equations in the same year, but Poincaré was content with the rigorous derivation, while Einstein went on looking for an explanation, for a mechanism. His famous example of the light bulb turned on in a moving train showed that length contraction is reasonable – even more, inescapable. Powerful as it is, Einstein's desire to find a mechanism for everything is very hard

to reconcile with the experimental demonstration of non-locality – of course, we have not the slightest idea of how Einstein would have reacted to Aspect's experiment.

75. Not all conceivable relationships are the object of physics. For instance, if there is a God, the world's relationship to Him is not accessible to physics, because we cannot modify this relationship and then observe the consequences of the changes.
76. This vision has often been called the *Copenhagen interpretation*, because Bohr worked there.
77. Let's look at one of the many problems of collapse. Consider the apparatus for two-particle correlations, and suppose collapse is really happening. How would one describe the measurement? Initially, one would say that the first particle to reach its measurement apparatus collapses into one or another detector, then informs the other one, which, when being measured itself, collapses into the suitable detector on its side. In such a description, one is forced to assume the problematic feature of superluminal hidden communication. But this is a minor point compared to another one: according to relativity, if two events are separated so that no light signal can join them, then time-ordering is not defined – in short, there is no way of saying which was the *first* particle to reach the detector. Arguments like this one are strong but inconclusive evidence against collapse, since some clever collapse models have been developed, for instance by GianCarlo Ghirardi.
78. H. Everett, *Rev. Mod. Phys.* **29**, 454 (1957).
79. Technically, the state of the universe now exhibits entanglement between the state of the particle and the state of the detector, physicist, etc.
80. I would like to draw the attention of the expert reader to the fact that the idea of many universes, or the 'multiverse', has also appeared in the context of cosmology, with a different flavour: if the Big Bang is just a fluctuation of some more fundamental field, then in this case, other fluctuations may be 'simultaneously' existing, and these would be parallel universes. For a simple and clear account of this question, see: M. Rees, *Just Six Numbers* (Basic Books, New York, 2001).
81. They read: the lattice of properties must satisfy the conditions of weak modularity and must possess the covering law.
82. Hence, for those who know relativity, a preferred frame (or more generally, a preferred foliation of space-time in equal-time slices) must

be assumed for quantum phenomena if one accepts the pilot-wave theory. In fact, the pilot waves of De Broglie and Bohm constitute a *quantum version of the ether*, the hypothetical support for light that Einstein's theory of relativity has declared useless. See chapter 12 of: D. Bohm, B.J. Hiley, *The Undivided Universe* (Routledge, London, 1993).

83. At the end of the paragraph devoted to unorthodox interpretations, I cannot refrain from using a perfectly out-of-context quote from C.S. Lewis, *That Hideous Strength* (Simon and Schuster paperback edition, New York, 1996), page 72. There, the hesitations of Jane Studdock in taking up her destiny are summarized by the sentence 'To avoid entanglements and interferences had long been one of her first principles.' Even though the words are used in their common English meaning, this sentence couldn't have gone unnoticed by a quantum physicist! It is a beautiful description of the goals of unorthodox interpretations.

84. To have an idea of the current situation of the interpretations as it is experienced in the world of physics, a slice of life is perhaps worth more than the tracts. Therefore I recommend the reader pay attention to the following story. In its March and April 1998 editions, *Physics Today* publishes an article in two parts written by Sheldon Goldstein. The article bears the title *Quantum Theory without observers.* In it, Goldstein presents a survey of the works about the interpretation of Bohm's pilot waves. It provokes a general outcry! In the February 1999 edition, *Physics Today* publishes letters from numerous very well known physicists expressing their disagreement with Goldstein's position. In the August 1999 edition, the review opens its doors to Robert Griffiths and Roland Omnès, to present their very much more orthodox interpretation 'of the coherent histories'.

85. I borrow this expression from: A. Shimony, *Conceptual Foundations of Quantum Mechanics*, in: P. Davies (ed.), *The New Physics*, (Cambridge University Press, Cambridge, 1989)

86. It is tempting to jump to the conclusion that these two forms of indeterminism are identical, and some people do. Specifically, some people conclude that Nature makes 'conscious' choices – or, in a slightly different version, that some intelligent Being is at work each time a quantum measurement is performed. Others believe that

quantum indeterminism is the physical substrate of human freedom. Such visions can neither be enforced nor falsified – as for their consistency, I leave that to the epistemological debate.

87. On this topic, I suggest: L. Smolin, *Three Roads to Quantum Gravity* (Basic Books, New York, 2002); B. Greene, *The Fabric of the Cosmos* (Alfred A. Knopf, New York, 2004). The reader must be warned that these are fascinating, but largely open problems even as mathematical theories, let alone as descriptions of reality. My impression from outside the field (gained, however, through discussion with the specialists) is that the hope, which is periodically raised by any partial advance, is thrown down after a few months or years of closer analysis. The above-mentioned books provide a rather equilibrate balance, although obviously the authors stress feats rather than failures.
88. For those who have studied quantum physics, I have listed more arguments supporting this claim in the proceedings of a conference, available online: xxx.lanl.gov/abs/quant-ph/0309113.
89. According to the most widespread vision of physics, both classical and quantum, any process could, in principle, be reversed; but since *we* have lost track of details, we are not able to reverse it. Thus, irreversibility is considered a practical consequence of our limited knowledge and control over Nature, not as a fundamental feature of it. The reader can't miss the strong implications of such a view, if extrapolated to truly everything (just imagine the possibility of reversing the processes of ageing and death).
90. Let me justify this preference for Bohr's over Everett's vision. Many people simply find Everett's vision 'crazy'. A more sophisticated argument consists in invoking Ockham's razor – against which Everett and his many worlds obviously stand – or in saying that Everett's is an undue extrapolation. I basically agree with all these reasons, but in my opinion none is very strong: crazy ideas have already happened in physics, the Creator needs not have shaped the world using Ockham's razor, and the conclusion of an undue extrapolation may be found to be true. The following argument is in my opinion stronger (but the reader may disagree). Note first that *Everett's vision is ultimately deterministic.* It is not deterministic in the usual sense: in any of the worlds, the results of measurements will always appear as objectively random (hence this interpretation is orthodox). But this vision states that our world and any of the other worlds

is emerging from an underlying quantum reality, whose evolution is deterministic – the non-expert reader may not see this, please believe me; the expert reader should, because any unitary quantum evolution (without measurement *à la* Bohr) is deterministic. Now, here is the point: if I had to commit to a deterministic worldview, then it is much simpler and more reasonable to adopt the strict deterministic view described in the main text, according to which everything (including quantum phenomena) is pre-established as in a screenplay – basically all the philosophical consequences (humans are only a part of a whole, our freedom is illusory, etc.) and definitely all the physical predictions are the same as in Everett's vision.

91. For a didactic presentation of Bose–Einstein condensates, visit the website 'Physics 2000' located at Boulder (http://www.colorado.edu/physics/2000/index.pl), section The Atomic Lab.
92. In famous fiction movies, it is indeed *matter* that is teleported. In Star Trek, a machine is needed only in one location, which can teleport to (or back from) any point of the universe; in 'The Fly', machines are needed in both locations. In this last example, however, matter is 'encoded' in a signal, which is then transmitted through a cable and 'decoded': since the signal travels through all intermediate locations, this is not strictly speaking a teleportation. I leave to experts of Star Trek the discussion, whether 'beaming' is supposed to work on a similar principle (without cable) or not.
93. The reader who knows quantum physics is aware that there is no state for which two spins 1/2 are parallel according to all directions; while there is a state, the 'singlet', in which two spins 1/2 are anti-parallel according to all directions. But for the present discussion, this detail does not matter.
94. Here are a few milestones. The protocol of teleportation was discovered in 1993: C.H. Bennett, G. Brassard, C. Crépeau, R. Jozsa, A. Peres, W.K. Wootters, *Phys. Rev. Lett.* **70**, 1895 (1993). The first two experiments were performed simultaneously in Rome and in Innsbruck: D. Boschi, F. Branca, F. De Martini, L. Hardy, S. Popescu, *Phys. Rev. Lett.* **80**, 1121 (1998); D. Bouwmeester, J.W. Pan, M. Eibl, K. Mattle, H. Weinfurter, A. Zeilinger, *Nature* **390**, 575 (1997). The first experiment where all four outcomes of the Bell measurement could be recorded was: Y.-H. Kim, S.P. Kulik, Y. Shih, *Phys. Rev. Lett.* **86**, 1370 (2001). For long-distance experiments, you should by now

know where to look: I. Marcikic, H. de Riedmatten, W. Tittel, H. Zbinden, N. Gisin, *Nature* **421**, 509 (2003). The Bell measurement and the analysis of C were done in two laboratories located some 50 metres apart from one another.

95. The two leading groups in ion traps, in Innsbruck and Boulder, achieved the feat simultaneously and their respective articles appeared in the same issue of Nature: M. Riebe *et al.*, *Nature* **429**, 734 (2004); M.D. Barrett *et al.*, *Nature* **429**, 737 (2004).
96. T.S. Eliot, *Four Quartets* (Faber and Faber, London, 1959): The Dry Salvages, end of movement III. By the way, the title of this last chapter was also borrowed from the same work of poetry (last verse of East Coker).

Index